Residential Construction Performance Guidelines

for Professional Builders & Remodelers

Third Edition

NATIONAL ASSOCIATION OF HOME BUILDERS · BOOKS THAT BUILD YOUR BUSINESS

NAHB
NATIONAL ASSOCIATION
OF HOME BUILDERS

Remodelors™ Council
Single Family Small Volume
Builders Committee

RESIDENTIAL
CONSTRUCTION
PERFORMANCE
GUIDELINES

RC
PG

Residential Construction Performance Guidelines for Professional Builders & Remodelers, Third Edition

Jill Tunick	Project Manager
David Jaffe	Legal Review
Wil Heslop	Executive Director, Business Management Department
Christine B. Charlip	Publisher, BuilderBooks
Torrie Singletary	Production Editor
Granville Woodson	Cover Design
Goodway Graphics of Virginia	Composition and Printing
Gerald M. Howard	NAHB Executive Vice President and CEO
Mark Pursell	NAHB Senior Staff Vice President, Marketing & Sales Group
Lakisha Campbell	NAHB Staff Vice President, Publications & Affinity Programs

This publication is designed to provide accurate and authoritative information in regard to the subject matter covered. It is sold with the understanding that the publisher is not engaged in rendering legal, accounting, or other professional service. If legal advice or other expert assistance is required, the services of a competent professional should be sought.

—From a Declaration of Principles jointly adopted by a Committee of the American Bar Association and a Committee of Publishers and Associations.

Printed in the United States of America
10 09 08 07 06 05   1 2 3 4 5
ISBN 0-86718-606-2
Library of Congress CIP information available on request.

For further information, please contact:

National Association of Home Builders
1201 15th Street, NW
Washington, DC 20005-2800
800-223-2665
Visit us online at www.builderbooks.com

Contents

Purpose of the Book

A Level of Expectation

Beyond building codes and local regulations, both contractors and their customers have long sought a measurable benchmark that deals with the expectations of <u>performance</u> in the goods and services provided by the residential construction industry. Although addressing matters of health, safety, and welfare are the mandated responsibility of those codes and regulations, a greater likelihood of matching the "other" dimension of a consumer's expectations will come from the acceptance of objective criteria regarding <u>performance</u>. Upon this premise, the prior two editions of *Residential Construction Performance Guidelines* were developed and refined to offer achievable minimum levels of workmanship for the products delivered.

It should be noted that the core of these criteria were first established as a basis of coverage under the <u>insured warranty program,</u> which was initially offered some 30 years ago. More than 20 years ago, the Remodelors Council of the National Association of Home Builders embraced a similar compilation of guidelines, which led to the published editions of *Quality Standards for the Professional Remodelor*. The joint effort of the Remodelors Council and the Single Family Small Volume Builders Committee culminated into the first edition in 1996. As such, many of the individual guidelines have remained as time-honored measures.

Currently, in certain geographical locations across the country, the *Residential Construction Performance Guidelines* have grown in acceptance to become the basis for evaluating performance by parties under residential construction contracts when dispute may lead to litigation or arbitration.

Third Edition Review

The performance guidelines in this manual were initially created and reviewed by more than 300 builders and remodelers. In addition, a number of specialty trade organizations participated in the review process and are listed in the appendix. This exhaustive review resulted in a first edition that builders and remodelers referenced with confidence as they communicated with their customers. The second and third editions are substantially similar to the initial publication. In both the second and the third editions, certain guidelines were updated to add clarity and thus make them more easily understood by both contractors and consumers alike. Certain issues were deleted while others were added to make the text more complete and comprehensive.

It remains an objective that the continued refinement of these guidelines can only lead to a mutual understanding of expectations by contractors and their customers. As such, the potential for disputes throughout the process can be greatly reduced. It is in that vein that the inclusion of an Applications section (see page xiii) and a Glossary of Common Terms (see page 99) was made. Additionally, performance guidelines were prepared where they were previously addressed as "none." Finally, any perceived ambiguities or redundancies were corrected as needed for clarity and comprehension.

Scope of the Construction Performance Guidelines

The guidelines are intended for use as a reference and should be interpreted with common sense. They should be applied only within the scope of the particular project being performed, and are not intended to answer all questions pertaining to workmanship of construction that might arise in the course of a typical residential construction project. The guidelines selected for this manual deal with those issues that most frequently give rise to questions for the contractor and the consumer. Although many contractors routinely build to tighter tolerances, this is a collection of minimum performance criteria and should be interpreted as such.

The developers of these guidelines examined typical building techniques and quality measures based on an overall view of residential construction within the United States. Local or regional differences in construction techniques and environmental conditions may make the application of some of these guidelines inappropriate.

Users of this manual should recognize that not all of these guidelines could be consistently applied across the United States. Common construction practice or the performance of materials or components under certain local conditions may prevent reasonable use of these guidelines for evaluation and suggested remedy. In these cases, the parties may want to expressly provide that some standard, other than the related guideline, will apply. Similarly, if the specific problem has resulted in locally developed guidelines, practice, or changes in the local building code, the parties may want the local guideline, practice, or code to take precedence over these guidelines.

The construction performance guidelines do not constitute a warranty nor are they intended as a substitute for a warranty; however, both parties may agree to incorporate them by reference within a warranty or within any other construction contract provision. They are separate and distinct from any manufacturers' or fabricators' warranties that may apply to materials and products used in the project.

Contractors often refer to these guidelines in the Dispute Resolution section of their contracts as the first step prior to any mediation, arbitration, or litigation. In essence, the use of these guidelines as the basic criteria has eliminated the need for any type of dispute resolution in a number of cases.

The use and application of these guidelines with regard to residential construction and remodeling work is strictly optional and at the discretion of the individual users, and if implemented, only refers to contractor-installed materials and services and not elements contracted by consumers and other trade contractors.

Nothing in this manual should be construed as policy, an endorsement, warranty (express or implied), or guaranty by the National Association of Home Builders or any persons or organizations involved in the creation of this manual of any technical descriptions, systems, details, requirements, materials, or products.

The National Association of Home Builders expressly disclaims any responsibility for any damage arising from the use, application, or reliance on the recommendations and information contained in this manual.

Scope of Responsibilities

Typically, numerous parties are involved in a residential construction project, whether it is building a new home or remodeling an existing one. Each of these parties has specific responsibilities to fulfill. The **Contract Documents** should provide a clear statement of the agreement between the contractor and the customer. In addition to the specific provisions of any contract, the following general responsibilities should be noted:

The Contractor: For the purposes of this book, the contractor is the entity named in the contract that has primary responsibility for completing the project. The contractor often employs others to assist him or her. In most cases, the contractor is responsible for all work assigned in the contract regardless of who actually performs the work. If the contractor is acting in a special role (for instance, as a construction manager), or the consumer selects others to work on the project who are outside the contractor's control, then the responsibility for evaluation and remedy of proposed problems may fall to other parties.

The Consumer: The consumer is the buyer of the product or service named in the contract; as such, the consumer is responsible for carefully reviewing the contract to ensure it accurately represents his or her expectations for the final product. Once the consumer accepts the project and moves into the home or occupies the newly renovated space, then he or she is responsible for routine maintenance and upkeep. Homes require a certain amount of care that is generally the consumer's responsibility. Additionally, consumers should note that in some of the guidelines contained in this publication, the contractor is not obligated to make repairs to items that fall within the consumer's maintenance responsibilities.

Manufacturer or Fabricator: Manufacturers and fabricators warrant many residential construction components that may fall outside the scope of the contractor's responsibilities, such as kitchen appliances, furnaces, air conditioners, and the like. Other less obvious items may include certain types of siding, roofing, or flooring. If there is a warranty question with one of these components, the consumer should be aware that the contractor might not be responsible for the performance of the product once it is installed. If a problem occurs, the consumer will often deal directly with the manufacturer or fabricator to have the problem evaluated and, if necessary, rectified unless otherwise specified in a contract. The contractor's responsibilities may end once he or she provides the appropriate information on how to contact the manufacturer or fabricator, unless otherwise specified in the contract.

Remodeling Projects

Remodeling is the process of expanding or enhancing an existing structure. There are inherent difficulties in melding the new and old in a way that meets the consumer's needs and is also aesthetically pleasing. Therefore, there are circumstances that call for the suspension of the application of these guidelines in order for the remodeling project to be successfully completed. These include, but are not limited to: the meeting of old, out-of-plumb or out-of-level structures with new structures; the appearance of new materials near weathered, existing materials; and the practical considerations for new projects to work within the limitations of existing buildings.

Because of the unique challenges of joining new with old, a remodeling contractor may build part of or the entire project outside the scope of these guidelines to achieve the contract objectives. When it is reasonable, the contractor may note and discuss a condition with the consumer before construction. It is also normal for a contractor (in the course of construction) to discover and accommodate conditions in the old structure that require solutions different from those the guidelines suggest. In these circumstances, the governing factor is meeting the needs of the consumer as outlined in the contract and complying with the local building code. **Note:** *Remodeling-specific items are in italics.*

How to Use This Manual

This manual is divided into chapters generally organized according to the usual sequence of events in the construction process. Nearly every chapter has major categories or sections; some also have smaller subsections. Each chapter contains individual construction performance guidelines.

The guidelines in this book are numbered according to the following sequence:

Chapter Number–Section Number–Guideline Number

Please note that the guideline numbering restarts under each new section within a chapter. Smaller subsections within chapters do not affect the numbering system.

Each construction performance guideline has three parts, with an optional fourth part:

Observation: A description of a particular construction condition.

Performance Guideline: The specific criterion for acceptable workmanship.

Corrective Measure: A description of the work required by the contractor to meet the performance guideline and/or the consumer's maintenance responsibility.

Discussion: (optional) An explanation of unique factors pertinent to the observation, performance guideline, or corrective measure.

The guidelines are supplemented by a glossary, a list of trade associations that have contributed to the guidelines in this manual, and an alphabetical subject index. The subject index is a comprehensive listing of applicable guidelines. Most guidelines are referenced by several listings that generally capture both contractor terminology and a typical consumer's description of a condition.

General Instructions

In many areas, construction is covered by a process that requires all work to be done in compliance with locally approved, applicable building and related codes or locally approved or adopted guidelines.

If any conflict arises between these guidelines and applicable requirements of locally approved codes or locally approved or accepted guidelines, as a matter of law, the code requirements or performance criteria may take precedence over these guidelines.

These performance guidelines apply only to contracting work as specified in the contract documents for the project. They do not apply to designs, plans, materials, or workmanship that is supplied by the consumer or is outside the scope of the particular project. They are also designed to apply only to the part of the job addressed in each guideline.

Explanation of Terminology in This Manual

Substantial completion of the project. A project has met substantial completion where the areas are functional for their intended use as stated by the contract (except for items noted prior to final presentation), and clean-up on the site has been completed.

Warranty period is defined as the duration of the applicable warranty provided by the contractor or any other period agreed to by the parties.

How to Incorporate These Guidelines Into a Warranty or Dispute Resolution Program

The warranty, like the contract, should clearly express the intent of the parties. The limited warranty describes the problems for which the contractor will be responsible after completion of the project, and specifies the time period during which the warranty is in force. Moreover, if a builder or remodeler warrants workmanship and materials in a warranty, he or she will want to provide some means of determining whether he or she is complying with the terms of the warranty. Without guidelines referenced, the parties run the risk of having to follow specified dispute resolution procedures where an arbitrary standard may be imposed.

Accordingly, the contract and/or warranty might include a statement such as the one that follows.

> All workmanship shall conform to the guidelines found in the publication *Residential Construction Performance Guidelines for Professional Builders and Remodelers*, National Association of Home Builders, 2005. If an item is not covered in that publication, standard industry practice shall govern. This may include the dispute resolution process as specified in the contract documents or by applicable laws.

If there are particular guidelines within this publication that the contractor or consumer does not feel are reasonable, they should be specifically excluded from all warranty or contract documents. Likewise, if there are particular guidelines that are not addressed, then by agreement, the contractor and consumer should add these to be part of the warranty and/or contract documents as referenced.

Other Ways to Use These Guidelines

- Make the guidelines available to consumers to help them understand the construction process, whether or not they are referenced in the contract.

- Avoid disputes by referencing an objective set of guidelines with third-party credibility.

- Make the guidelines available to mediators, arbitrators, and judges to help them understand the acceptable performance criteria.

- Make the guidelines available to building code officials so that they can distinguish them from code compliance issues.

- Make the guidelines available to trade contractors whose profession is an integral part of the construction process.

- Make the guidelines available to city, county, and state officials to consider adoption of the guidelines as their accepted criteria.

- Make the guidelines available to private, third-party inspectors and their trade associations to facilitate their compliance if designated as a condition of acceptable fit or finish.

Applications

Through the years, some methods of application have evolved to more readily assist both contractors and consumers in making accurate measurements of such items as ridges, cracks, gaps, lippage, or variations in plumbness or levelness. Because the majority of these are generally less than a 1-inch dimension, the use of the width of coins is suggested to approximate the measurement of variation. For example, the width of a U.S. dime approximates 1/32nd of an inch while a U.S. quarter is reasonably close to 1/16th of an inch.

With this in mind, measurements can be made using multiples of coins to make fairly accurate measurements to determine compliance with the stated tolerances, as well as to determine the extent of repairs needed where the element exceeds the guideline.

A second suggestion refers to dimensions of levelness of surfaces, or those within horizontal measurement. While a standard 2-foot or 4-foot "carpenter's level" is readily available at most hardware or home stores, the search for a standard 32-inch level has been unsuccessful. To that end, removing 8 inches from both ends of a fiberglass 4-foot level will yield a fairly accurate tool for this need.

For Additional Information

Users are encouraged to provide comments and suggestions regarding their experiences with the guidelines. This could include applications that have been found useful in determining compliance. Responses should refer by name to *Residential Construction Performance Guidelines, Third Edition* and should be submitted in writing to NAHB Business Management Department, 1201 15th Street NW, Washington, DC 20005. Your comments will be considered in the preparation of future editions of this book.

BuilderBooks.com, a division of the National Association of Home Builders, has several publications available to assist with the development of contract documents, a comprehensive warranty program, and a variety of construction and customer service-related topics. There are also resources created specifically for the new-home buyer or remodeling customer. For a catalog and more information, call BuilderBooks at 1-800-223-2665. Visit BuilderBooks online at www.builderbooks.com.

Visit www.nahb.org/biztools and log on as an NAHB member for a variety of business management resources to help you work more profitably and productively. You'll find articles, books, and educational courses organized by category. Whether you need help with accounting and financial management, business and strategic planning, computers and information, customer service, the regulatory environment and your business, construction management, human resources, or sales and marketing, we have information for you. Visit and bookmark www.nahb.org/biztools.

Acknowledgments

NAHB gratefully acknowledges the leadership and commitment of the two groups that worked together to produce *Residential Construction Performance Guidelines, Third Edition:*

The NAHB Remodelors™ Council
Douglas Sutton, Sr., CGR, CAPS, Sutton Siding & Remodeling, Inc., Chair

The Single Family Small Volume Builders Committee
John H. Moffitt, Jr., Coldwell Banker John Moffitt and Associates; Moffitt Development Company, Inc., Chair

We thank Work Group Chairman John H. Moffitt, Jr., for directing the production of *Residential Construction Performance Guidelines, Third Edition.* We also thank the Work Group Leaders and Members listed below. This book would not have been possible without their dedication and expertise.

Work Group Leaders:
    Robert Bell, CGR, CAPS, Bell's Remodeling
    Daimon Doyle, GMB, CAPS, Doyle Custom Homes, Inc.
    Robert T. Merz, Construction Arbitration Associates, Ltd.
    Richard G. Reynolds, R.G. Reynolds Homes, Inc.
    J. Bradley Simons, Cottage Homes Corporation

Work Group Members:
    Charles Austin, Austin Signature Homes
    Steve Begshaw, Home Builders Association of Greater Kansas City
    Bob Blayden, CGR, CAPS, Blayden Design/Build
    Fred Buckley, Home Builders Association of Greater Kansas City
    Jeffrey Carpenter, Monticello Homes, Inc.
    Ed Carr, CGB, Comfort Home Corporation
    Jerry Clark, Clark & Son Construction, Inc.
    Terry C. Collins, Portland Cement Association
    Lawrence M. Cramer, L. Cramer Designers & Builders
    Matt Dobson, Vinyl Siding Institute, Inc.
    Wayne Foley, W.M. Foley Construction Corporation
    Don Garramone, Garramone Builders, Inc.
    Alan Gelet, Englemann, Inc.
    John A. Gill, Quality Builders Warranty Corporation
    Alan Hanbury, Jr., CGR, CAPS, House of Hanbury Builders, Inc.
    Robert Hanbury, CGR, House of Hanbury Builders, Inc.
    Dave Holtwick, Home Builders Association of Greater Kansas City
    Bob Hubbard, R.C. Hubbard Construction, Inc.

John Hubbard, 2-10 Home Buyers Warranty, Inc.
Mike J. Kegley, B.O.L.D. Homes, Inc.
Sylvia Kellogg, USG Corporation
Steve Lains, Builders Association of Greater Indianapolis
Richard Matzen, American Society of Home Inspectors
Peter Merrill, Construction Dispute Resolution Services, LLC
Paul Michelsohn, Jr., Michelsohn & Daughter Construction, Inc.
Greg Miedema, CGR, CAPS, CGB, Dakota Builders, Inc.
Mike Moran, Simpson Strong-Tie
Richard Morin, Kara Elizabeth Homes
David Nielsen, Home Builders Association of Metro Portland
John J. Piazza, Sr., Piazza Construction, Inc.
Barry Pollock, Bonded Builders Home Warranty Association
Donald L. Pratt, Wake-Pratt Construction Company
Jeffrey Prostor, Brookfield Homes
Dwight "Sonny" Richardson, Jr., Richardson Home Builders, Inc.
Stephen L. Robinson, CGR, CAPS, CGB, GMB, RE Construction
Roddy O. Sample, CGR, Roddy O. Sample & Associates, Inc.
George Schluter, CGB, GWS, Inc.
Glenn Singer, CertainTeed Corporation
Jim Thomas, J.L. Thomas Construction
Ray Tonjes, Ray Tonjes Builder, Inc.
Mike Turner, CGR, The Home Service Store
Michael H. Weber, Portland Cement Association
M.M. "Mike" Weiss, CGR, CAPS, CGB, GMB, Weiss & Company, Inc.
Duane P. Willenbring, CGB, Willenbring Construction, Inc.

We acknowledge with appreciation the contributions of the following NAHB
Standing Committees:
Building Product Issues Committee
Business Management & Information Technology Committee
Commercial Builders Council
Construction, Codes & Standards Committee
Construction Safety & Health Committee
Credentials Committee
Custom Home Builders Committee
Multifamily Council
Research Committee
Resolutions Committee
Single Family Production Builders Committee
Single Family Small Volume Builders Committee
State & Local Government Affairs Committee

Our thanks to the following organizations for sharing review materials with
us:
Bonded Builders Home Warranty Association
Builders Association of Greater Indianapolis

Building Industry Association of Washington
Canadian Home Builders Association
Greater Atlanta Home Builders Association
Home Buyers Warranty Corporation
MacLellan Wolfson Associates
Metropolitan Builders Association of Greater Milwaukee
Michigan Association of Home Builders
Portland Cement Association
Quality Builders Warranty Corporation
Southern Nevada Home Builders Association
Texas Association of Builders
Texas Residential Construction Commission
Tile Council of America
2-10 Home Buyers Warranty, Inc.
U.S. Department of Housing and Urban Development

We also wish to thank John Kubik of Inspection Services Group, Inc., for giving NAHB permission to use definitions from ISGI's Glossary of Building Terminology in the *Residential Construction Performance Guidelines, Third Edition* Glossary of Common Terms.

In addition, NAHB gratefully acknowledges the leadership and commitment of the two groups that worked together to produce *Residential Construction Performance Guidelines, Second Edition*:

The NAHB Remodelors™ Council
Janet Williams, Williams-Builder, Chair

The Single Family Small Volume Builders Committee
Nelson Vaughan, Vaughan & Sons Builders, Inc., Chair

We would like to acknowledge the hard-working and dedicated Work Group Leader who directed the meticulous review of the guidelines:

Robert T. Merz, Construction Arbitration Associates, Ltd.

And the Work Group Members:
Robert Bell, CGR, CAPS, Bell's Remodeling
John Barrows, Telemark Construction, Inc.
Dan Bawden, CGR, Legal Eagle Contractors Company
Charles Beatch, Beatch Construction
Bob Blayden, CGR, Blayden Design/Build
Shirley Blayden, Blayden Design/Build
William Fannin, Jr., Fannin Builders, Inc.
Michael Fawcett, Michael Fawcett Construction
Ivan Foerster, Foerster Design & Construction
Dwight Griffith, CGR, Griffith Brilhard Builders

Jeff Hager, Hager Construction
Jess Hall, Hall Quality Builders
Alan E. Hanbury, Jr., CGR, House of Hanbury Builders, Inc.
Robert D. Hanbury, CGR, House of Hanbury Builders, Inc.
Norm Hutchens, CGR, Norm Hutchens Builders, Inc.
Frank Jones, CGR, Cavalry Construction
Robert R. Jones, Robert R. Jones Associates
Richard Miller, Richard Miller Construction, Inc.
Douglas Nelson, New Spaces
Curt Ostrom, C.N. Ostrom & Sons, Inc.
Gretchen Palmer, Palmer Homes
Vernon Palmer, Palmer Homes
Michael F. Payne, Payne & Payne Builders, Inc.
John J. Piazza, Sr., Piazza Construction, Inc.
Avon Privette, Statesboro Builders
Walt Rebmann, Buck Construction Company
Clete Reinhart, Reinhart Construction
Chris Repp, CGR, Repp Construction Inc.
David R. Roberts, David R. Roberts Builders
Michael Schultz, Michael Schultz Construction
Mike Turner, CGR, T.L.C. Builders, Inc.
Nelson Vaughan, N. Vaughan & Sons, Inc.
Jonathan Wallick, CGR, Wallick Construction Company

This book would not exist today had it not been for the original outstanding commitment from many individuals and organizations. We are compelled to recognize the contributions of those who produced the first edition of *Residential Construction Performance Guidelines*. They include:

The NAHB Remodelors™ Council
Bill Asdal, Asdal & Co. Builders LLC, Chair

The Single Family Small Volume Builders Committee
Suzanne Grove, Grove Construction, Chair

Work Group Leaders:
Robert Bell, CGR, Bell's Remodeling
Norm Hutchens, CGR, Norm Hutchens Builders, Inc.
Robert T. Merz, Construction Arbitration Associates, Ltd.
John J. Piazza, Sr., Piazza Construction, Inc.

Work Group Members:
Dan Bawden, CGR, Legal Eagle Contractors Company
Charles Beatch, Beatch Construction
Bob Blayden, CGR, Blayden Design/Build
Shirley Blayden, Blayden Design/Build
James Christo, Christo Design/Builder

William Fannin, Jr., Fannin Builders, Inc.
Michael Fawcett, Michael Fawcett Construction
Dwight Griffith, CGR, Griffith Brilhard Builders
Jeff Hager, Hager Construction
Jess Hall, Hall Quality Builders
Alan E. Hanbury, Jr., CGR, House of Hanbury Builders, Inc.
Robert D. Hanbury, CGR, House of Hanbury Builders, Inc.
Frank Jones, CGR, Cavalry Construction
Robert R. Jones, Robert R. Jones Associates
Bob Merriman, New Rooms & Spaces
Richard Miller, Richard Miller Construction, Inc.
Curt Ostrom, C.N. Ostrom & Sons, Inc.
Fred Parker, Fred Parker Company, Inc.
Michael F. Payne, Payne & Payne Builders, Inc.
Kim Post, Newell Construction, Inc.
Avon Privette, Statesboro Builders
Walt Rebmann, Buck Construction Company
Clete Reinhart, Reinhart Construction
Chris Repp, CGR, Repp Construction, Inc.
David R. Roberts, David R. Roberts Builders
Mike Turner, CGR, T.L.C. Builders, Inc.
Nelson Vaughan, N. Vaughan & Sons, Inc.
Jonathan Wallick, CGR, Wallick Construction Company

We acknowledge with appreciation the contributions of:

NAHB Standing Committees:
Construction and Codes Committee
Single Family Custom Builders Committee
Single Family Production Builders Committee

State and Local Associations' Boards of Directors and Work Groups:
Blue Ridge Home Builders Association, VA
Builders Association of Greater Boston, MA
Builders Association of the Twin Cities, MN
Building Industry Association of Central Coast, CA
Building Industry Association of Central Ohio
Building Industry Association of Southeastern Michigan
Greater Atlanta Home Builders Association, GA
Greater Houston Builders Association, TX
Home Builders Association of Greater Chicago, IL
Home Builders Association of Greater Fox Valley, IL
Home Builders Association of Greater Kalamazoo, MI
Home Builders Association of Maryland
Home Builders Association of Metro Denver, CO
Indiana Home Builders Association
Master Builders Association of King & Snohomish Counties, WA

New Jersey Builders Association
Santa Barbara Contractors Association, CA
Skagit-Island Counties Builders Association, WA

Individual Reviewers:

William Allen	Bruce Giffin	William Pierce
David Anderson	Daniel Green	Steve Pjesky
Marvin Ausherman	Dale Hallen	Kim Post
Judy Barnes	Jeff Hansell	Lowell Pratt
Carol Baye	Kyle Harder	Matt Robb
Randy Birdwell	Charlene Hatlen	Ross Robbins
David Bock	John Hayman	David Roberts
Sam Bradley	Dennis Henderlong	Barry Rutenberg
Richard Brown	Louise Henson	Richard Salm
Scott Brown	Michael Himes	Dwight Sandlin
Thomas Bruin	Lee Holztrager	Rod Saponaro
Ron Burton	Ken Hubers	Jim Sattler
Scott Campbell	Michael Jackson	Tom Sattler
Pat Carmichael	Alan Jaffe	Bill Sherman
Andy Cattano	David Jaffe	David Showers
Tom Cauble	Clark Johnson	Robert Simmons
Howard Chilcutt	Ron Jones	Michael Sivage
Bill Clayburg	Paul Joyal	Larry Skelton
John Cochenour	John Kavanagh	John Small
Ken Coe	Dennis Kelly	Carol Smith
Scott Colwell	Bob Kelzer	Steve Spanjer
Calvin Cozort	Aaron Kolkey	Ken Staples
Larry Cramer	Gary Lewis	Tom Stephani
Jean Delaura	Roy Lund	Jim Stump
Manny Dembs	Tim Lund	Bill Styczynski
Ray Derby	Pierre Mainguene	Richard Sullivan
Steve Dixon	Richard Mather	Curt Swanson
Edward Drake	Dean McFarland	Allen Tajmir
Tom Drew	Dave McGrath	Nancy Taylor
Angela Marie Driver	Anne McKnight	Chip Vaughn
Richard Durling	Joe McMahon	Bill Wall
Stan Dveris	Ron McOmber	Keith Waters
Bradley Elder	Scott Miller	Bob Wickman
Art Elliott	Tom Molidar	Dennis Wilson
Bob Elliott	Dean Mon	Steve Wilson
Joe Ernst	John Morel	Dennis Wiltgen
Michael Fink	Roy Nash	
Ivan Foerster	John Osborn	
Wayne Foley	Jack O'Shura	
Mike Gaffney	Vernon Palmer	
Norman Gerber	Isreal Pena	

Site Work

1-0-1 **Observation:** The ground has settled around the foundation, over utility trenches, or in other areas.

Performance Guideline: Settling of ground around foundation walls, over utility trenches, or in other filled areas shall not interfere with water drainage away from the home.

Corrective Measure: If the contractor provided final grading, one time only the contractor will fill areas that settle more than 6 inches and that affect proper drainage. The consumer will be responsible for removal and replacement of shrubs, grass, other landscaping, pavement, sidewalks, or other improvements affected by placement of such fill.

1-0-2 **Observation:** The site does not drain properly.

Performance Guideline: To ensure proper drainage in the immediate area around the home, the contractor shall establish the necessary grades and swales if the work is included in the contract. Standing water or ponds of water shall not remain for extended periods in the immediate area of the house after a rain (generally no more than 24 hours), except in swales that drain other areas or in areas where sump pumps receive discharge. In these areas a longer period can be anticipated (generally no more than 48 hours). Water may stand longer during periods of heavy rains, especially when heavy rains occur on successive days. No grading determination shall be made while frost or snow is on the ground or while the ground is saturated.

Corrective Measure: If grading is part of the construction or purchase agreement, the contractor is responsible for initially establishing the proper grades and swales.

Discussion: Grass and other landscaping are integral components of the storm water management practice needed to minimize erosion from the site. It is the consumer's responsibility to maintain such grass and other landscaping to help ensure proper functioning of the site drainage system. The consumer is responsible for maintaining such grades and swales once the contractor has properly established them.

1–0–3   **Observation:** The site has soil erosion.

Performance Guideline: The contractor is not responsible for soil erosion due to acts of God, or other conditions beyond the contractor's control.

Corrective Measure: No action is required. The contractor is not responsible for erosion due to acts of God, exceptional weather conditions, site alterations by the consumer, lack of maintenance by the consumer, or other conditions beyond the contractor's control.

1–0–4   **Observation:** Water from a nearby or adjacent property flows onto the consumer's lot.

Performance Guideline: The contractor is responsible for providing a reasonable means of draining off the lot water that is created (rain, melting snow or ice) on the lot, but is not responsible for water flowing from a nearby or adjacent property or on which no dwelling has been erected other than providing proper slopes around the newly erected dwelling.

Corrective Measure: It is the contractor's responsibility to control water only in the immediate area of the new dwelling.

1–0–5   **Observation:** Existing trees, shrubs, or other vegetation may be damaged in the course of construction.

Performance Guideline: The contractor will review the existing condition of the landscape with the consumer. The contractor will make a reasonable and cost-effective effort to preserve existing landscaping, but the survival of existing landscaping cannot be guaranteed.

Corrective Measure: No contractor action is needed.

Foundation

General

2–1–1 **Observation:** The foundation is out of square.

Performance Guideline: As measured at the top of the foundation wall, the diagonal of a triangle with sides of 12 feet and 16 feet shall be no more than 1 inch more or less than 20 feet. *Remodeling Specific: A contractor and consumer may agree to build an addition out of square in order to keep a new exterior wall on line with an existing wall of an out-of-square house.*

Corrective Measure: The contractor will make necessary modifications to the foundation not complying with the performance guidelines for squareness to provide a satisfactory appearance. The contractor may square the first-floor deck or walls by cantilevering over the foundation or locating the deck or walls inset from the outside face of the foundation.

Discussion: Squareness is primarily an aesthetic consideration. The corrective measure emphasizes the primarily aesthetic nature of squareness and makes the criterion for correction "a satisfactory appearance." This allows the contractor to make either a structural change or some cosmetic modification as most appropriate.

There are many instances in which the squareness of a foundation is not of consequence because subsequent construction provides an opportunity to make corrections.

2–1–2 **Observation:** The foundation is not level.

Performance Guideline: This guideline applies only when the levelness of the foundation adversely impacts subsequent construction. As measured at the top of the foundation wall, no point shall be more than 1/2 inch higher or lower than any point within 20 feet. *Remodeling Specific: The contractor and the consumer may agree to build an addition out of level in order to keep the floor of an addition on the same plane, and the roof ridge on the same line, as those of an existing, out-of-level structure.*

Corrective Measure: The contractor will make necessary modifications to any part of the foundation or to subsequent construction to meet the performance guideline for levelness. This can be effected by leveling the sills with shims, mortar, appropriate fillers, or other methods.

Discussion: There are many instances in which the levelness of a foundation is not of consequence because subsequent construction provides an opportunity to make corrections.

Levelness is both an aesthetic and functional consideration. Out-of-level floors can cause "stair stepping" of 4x8-foot sheathing, siding, paneling, and cabinets, and square walls must be "racked" into parallelograms when plumbing is installed. Liquids can run off countertops, and, in extreme cases, consumers will perceive that they are walking uphill or downhill. The contractor and the consumer may agree to build an addition out of level in order to keep the floor of an addition on the same plane, and/or the roof ridge on the same line, as those of an existing, out-of-level structure.

2–1–3   **Observation:** There is a crack in a concrete footing.

Performance Guideline: Cracks greater than 1/4-inch in width are considered excessive.

Corrective Measure: The contractor shall repair any cracks in excess of the performance guideline.

Interior Concrete Slab

2–2–1   **Observation:** A concrete slab within the structure has separated or moved at control (expansion and contraction) joints.

Performance Guideline: Concrete slabs within the structure are designed to move at control joints.

Corrective Measure: Because this is normal, no corrective action is required.

Discussion: Control joints are placed in concrete for the very purpose of encouraging cracking to take place at the joints instead of in random places.

2–2–2   **Observation:** Efflorescence is present on the surface of the basement floor.

Performance Guideline: This is a typical condition caused by moisture reacting with the soluble salts in concrete and forming harmless carbonate compounds.

Corrective Measure: Because efflorescence is a typical chemical reaction within concrete, no corrective measures are required of the contractor.

Discussion: Efflorescence is evidenced by the presence of a white film on the surface of the concrete. It is a particularly common occurrence where masonry or concrete are in contact with high moisture levels as may be found in basements.

2–2–3 **Observation:** The concrete floor or slab is uneven.

Performance Guideline: Except where the floor or portion of the floor has been designed for specific drainage purposes, concrete floors in living areas shall not have pits, depressions, or areas of unevenness exceeding 3/8-inch in 32 inches.

Corrective Measure: The contractor will correct or repair the floor to meet the performance guideline.

Discussion: A repair can be accomplished by leveling the surface with a material designed to repair uneven concrete.

2–2–4 **Observation:** The concrete floor slab is cracked.

Performance Guideline: Minor cracks in concrete floor slabs are normal. Cracks exceeding 3/16-inch in width or 3/16-inch in vertical displacement shall be repaired if the slab is in conditioned space or the crack interferes with the installation of finish flooring.

Corrective Measure: The contractor will repair cracks that do not meet the performance guideline.

Discussion: Repairs can be made by using a material designed to fill cracks in concrete.

2–2–5 **Observation:** Interior concrete work is pitting or spalling. Pitting is evidenced by concrete that has flaked or peeled from the outer surface. Spalling is evidenced by concrete that has chipped.

Performance Guideline: Interior concrete surfaces shall not pit or spall.

Corrective Measure: The contractor will repair defective concrete surfaces using materials designed for this purpose.

2–2–6 **Observation:** The interior concrete slab has a loose, sandy surface. This is called "dusting."

Performance Guideline: The surface shall not be so sandy as to cause a problem for the finish flooring to be applied.

Corrective Measure: The surface shall be corrected so as to be suitable for the finish flooring that the contractor had reason to anticipate would be applied.

Basement and Crawl Space Concrete Block Walls

2–3–1 **Observation:** A concrete block basement or crawl space wall is cracked.

Performance Guideline: Cracks in concrete block basement or crawl space walls shall not exceed 1/4-inch in width.

Corrective Measure: The contractor will repair cracks to meet the performance guideline.

Discussion: Shrinkage cracks are common in concrete block masonry and should be expected in crawl space and basement walls. Cracks may be vertical, diagonal, horizontal, or stepped-in masonry joints. Repairs can be made by using a material designed to fill cracks in concrete.

2–3–2 **Observation:** A concrete block basement wall is out of plumb.

Performance Guideline: Block concrete walls shall not be out of plumb greater than 1 inch in 8 feet when measured from the base to the top of the wall. *Remodeling Specific: If tying into an existing foundation that is out of plumb, the contractor and consumer will review the existing conditions and scope of work. The contractor will make a reasonable and cost-effective effort to meet the performance guideline while complying with the existing building code.*

Corrective Measure: The contractor shall repair any deficiencies in excess of the performance guideline. If the wall is to remain unfinished per contract, and the wall meets building codes as evidenced by passed inspections, then no corrective action is required.

2–3–3   **Observation:** A concrete block basement wall is bowed.

Performance Guideline: Concrete block walls shall not bow in excess of 1 inch in 8 feet.

Corrective Measure: The contractor shall repair any deficiencies in excess of the performance guideline. If the wall is to remain unfinished per contract, and the wall meets building codes as evidenced by passed inspections, then no corrective action is required.

Basement and Crawl Space Poured Walls

2–4–1   **Observation:** A poured concrete basement wall is out of plumb.

Performance Guideline: Finished concrete walls shall not be out of plumb greater than 1 inch in 8 feet when measured vertically. *Remodeling Specific: If tying into an existing foundation that is out of plumb, the contractor and consumer will review the existing conditions and scope of work. The contractor will make a reasonable and cost-effective effort to meet the performance guideline while complying with the existing building code.*

Corrective Measure: The contractor shall repair any deficiencies in excess of the performance guideline. If the wall is to remain unfinished per contract, and the wall meets building codes as evidenced by passed inspections, then no corrective action is required.

2–4–2   **Observation:** An exposed concrete wall has pits, surface voids, or similar imperfections in it.

Performance Guideline: Surface imperfections larger than 1 inch in diameter or 1 inch in depth are considered excessive.

Corrective Measure: The contractor will repair holes that do not meet the performance guideline.

Discussion: Pits, surface voids, and similar imperfections are sometimes called "bug holes." More technically, they are called "air surface voids" and are caused by air entrapped at the concrete and concrete form interface. The technical term for larger voids is "honeycomb" and must be dealt with in accordance with this guideline. One method of repair is to fill the hole or void with a suitable product. The repaired area is unlikely to match the color or texture of the surrounding concrete.

2-4-3   **Observation:** A poured concrete basement wall is bowed.

Performance Guideline: Concrete walls shall not bow in excess of 1 inch in 8 feet when measured from the base to the top of the wall.

Corrective Measure: The contractor shall repair any deficiencies in excess of the performance guideline. If the wall is to remain unfinished per contract, and the wall meets building codes as evidenced by passed inspections, then no corrective action is required.

2-4-4   **Observation:** A poured concrete basement or crawl space wall is cracked.

Performance Guideline: Cracks in poured walls shall not exceed 1/4-inch in width.

Corrective Measure: The contractor will repair cracks that do not meet the performance guideline.

Discussion: Shrinkage cracks and other cracks are common and are inherent in the drying process of poured concrete walls. They should be expected in these walls due to the nature of concrete. The only cracks considered under warranty claims are cracks that permit water penetration or horizontal cracks that cause a bow in the wall.

2-4-5   **Observation:** A cold joint is visible on exposed poured concrete foundation walls.

Performance Guideline: A cold joint is a visible joint that indicates where the pour terminated and continued. Cold joints are normal and should be expected to be visible. Cold joints should not be an actual separation or a crack that exceeds 1/4-inch in width.

Corrective Measure: The contractor will cosmetically repair any cold joint that exceeds 1/4-inch in width.

Moisture and Leaks

Basement Floor and Walls

2-5-1   **Observation:** Dampness is evident on basement walls or the floor.

Performance Guideline: Dampness caused by condensation of water vapor on cool walls and floors is not the responsibility of the contractor.

Corrective Measure: Dampness due to condensation is caused by high moisture content in the air. It is the consumer's responsibility to control the humidity.

Discussion: The consumer should maintain proper grade away from the dwelling.

2–5–2   **Observation:** The basement leaks.

Performance Guideline: Leaks resulting in actual trickling of water shall be repaired. Leaks caused by landscaping improperly installed by the consumer, or by the consumer's failure to maintain proper grades, are not the contractor's responsibility. New-construction walls and floors may become damp as concrete, mortar, and other materials dry. Dampness alone is not considered a deficiency.

Corrective Measure: The contractor will take such action as is necessary to correct basement leaks, except where the cause is determined to result from the consumer's actions or negligence.

Crawl Spaces

2–5–3   **Observation:** Water accumulates in the interior crawl space.

Performance Guideline: Crawl spaces should be graded and proper exterior foundation drains be provided to prevent water from accumulating deeper than 3/4-inch and greater than 9 square feet in the crawl space area.

Corrective Measure: The contractor will take corrective measures to meet the performance guideline. The contractor is not responsible if the exterior grading was provided by the consumer or the consumer failed to maintain grades established by the contractor.

2–5–4   **Observation:** Condensation is evident on the crawl space surface.

Performance Guideline: The contractor shall install the ventilation required by the prevailing building code.

Corrective Measure: If the crawl space is ventilated as required by

9

applicable building codes, then the contractor need make no further corrective actions. Further reduction of condensation is a consumer maintenance responsibility.

Discussion: Temporary conditions that cause condensation that cannot be eliminated by ventilation and a vapor barrier may include:

- Night air gradually cools the interior surfaces of the crawl space. In the morning, moisture picked up by sun-warmed air migrates into the crawl space and condenses on cool surfaces.
- At night, outside air may rapidly cool foundation walls and provide a cool surface on which moisture may condense.
- If the house is left unheated in the winter, floors and walls may provide cold surfaces on which moisture in the warmer crawl space air may condense.
- Excessive moisture inside a heated house may reach the dew point within or on the colder bottom surface of vapor-permeable floor insulation. Condensation can be reduced by placing a vapor barrier between the insulation and the floor sheathing. If condensation must be reduced substantially, the consumer can do so by sealing and dehumidifying or heating the crawl space, or by heating and dehumidifying the house.

Columns

2–6–1   **Observation:** An exposed wood column is bowed or is out of plumb.

Performance Guideline: Exposed wood columns shall not bow or be out of plumb more than 3/4-inch in 8 feet.

Corrective Measure: Exposed wood columns out of plumb in excess of 3/4-inch in 8 feet when measured vertically shall be replaced or repaired.

Discussion: Wood columns may become distorted as part of the drying process. Bows and other imperfections that develop after installation cannot be prevented or controlled by the contractor.

2–6–2   **Observation:** An exposed concrete column is installed bowed or out of plumb.

Performance Guideline: Exposed concrete columns shall not be installed with a bow in excess of 1 inch in 8 feet. They should not be installed out of plumb in excess of 1 inch in 8 feet.

Corrective Measure: The contractor shall repair any deficiencies in excess of the performance guideline.

2–6–3   **Observation:** A masonry column is out of plumb.

Performance Guideline: Masonry columns should not be constructed out of plumb in excess of 1 inch in 8 feet.

Corrective Measure: The contractor shall repair any deficiencies in excess of the performance guideline.

2–6–4   **Observation:** A steel column is out of plumb.

Performance Guideline: Steel columns shall not be out of plumb in excess of 3/8-inch in 8 feet when measured vertically.

Corrective Measure: The contractor shall repair any deficiencies in excess of the performance guideline.

Wood Floor Framing

Floor System

3–1–1   **Observation:** Springiness, bounce, shaking, or visible sag is present in the floor system.

Performance Guideline: All beams, joists, headers, and other structural members shall be sized according to the manufacturers' specifications or local building codes.

Corrective Measure: The contractor will reinforce or modify, as necessary, any member of the floor system not meeting the performance guideline.

Discussion: Deflection may indicate insufficient stiffness in the lumber, or may reflect an aesthetic consideration independent of the strength and safety requirements of the lumber. Structural members are required to meet standards for both stiffness and strength. When a consumer's preference is made known before construction, the contractor and the consumer may agree upon a higher standard.

Beams, Columns, and Posts

3–2–1   **Observation:** An exposed wood column or post is split.

Performance Guideline: Sawn wood columns or posts shall meet the grading standard for the species used. Splits that exceed 3/8-inch in width and more than 4 inches in length at time of installation or that develop during the warranty period are considered excessive.

Corrective Measure: The contractor will repair or replace any beam or post that does not meet the guideline. Filling splits is acceptable to have structural members meet the guideline.

Discussion: Columns and posts will sometimes split as they dry after installation. Splitting is acceptable and is not a structural concern if columns or posts have been sized according to manufacturers' specifications or local building codes. Splitting is primarily an aesthetic rather than a structural concern.

3–2–2   **Observation:** An exposed wood beam is split.

Performance Guideline: Sawn wood beams shall meet the grading standard for the species used. Splits that exceed 3/8-inch in width and 4 inches in length at time of installation or that develop during the warranty period are considered excessive.

Corrective Measure: The contractor will repair or replace any sawn wood beam that does not meet the guideline. Filling splits is acceptable to have structural members meet the guideline.

Discussion: Beams 2 1/2 inches or greater in thickness (which normally are not kiln dried) will sometimes split as they dry after installation. Splitting is acceptable and is not a structural concern if the sawn lumber beams have been sized according to manufacturers' specifications or local building codes. Splitting is primarily an aesthetic rather than a structural concern.

3–2–3   **Observation:** An exposed wood beam or post is twisted or bowed.

Performance Guideline: Exposed wood posts and beams shall meet the grading standard for the species used. Posts and beams with bows and twists exceeding 3/4-inch in an 8-foot section shall not be installed, and those that develop bows and twists exceeding 3/4-inch in an 8-foot section are considered excessive.

Corrective Measure: The contractor will repair or replace any beam or post with a bow or twist that exceeds the guideline.

Discussion: Beams and posts, especially those 3 1/2 inches or greater in thickness (which normally are not kiln dried) will sometimes twist or bow as they dry after milling or installation. Twisting or bowing is usually not a structural concern if posts and beams have been sized according to manufacturers' specifications or local building codes.

3–2–4   **Observation:** An exposed wood beam or post is cupped.

Performance Guideline: Cups exceeding 1/4-inch in 5 1/2 inches are considered excessive.

Corrective Measure: The contractor will repair or replace any beam or post with a defect that does not meet the guideline.

14

Discussion: Cupped lumber is lumber that has warped or cupped across the grain in a concave or convex shape. Beams and posts, especially those 3 1/2 inches or greater in thickness (which normally are not kiln dried), will sometimes cup as they dry after milling or installation.

Plywood and Joists

3–3–1   **Observation:** The wood floor squeaks or the subfloor appears loose.

Performance Guideline: Squeaks caused by a loose subfloor are unacceptable, but totally squeak-proof floors cannot be guaranteed.

Corrective Measure: The contractor will refasten any loose subfloor or take other corrective action to attempt to reduce squeaking to the extent possible within reasonable repair capability without removing floor or ceiling finishes.

Discussion: There are many possible causes of floor squeaks. One of the more common sources of squeaks is wood moving along the shank of a nail. Squeaking frequently occurs when lumber, plywood, or boards move slightly when someone walks over them. Boards and plywood may become loose due to shrinkage of the floor structure or subfloor as it dries after installation or seasonal changes in temperature and humidity. Nails used to fasten metal connectors (joist hangers, tie-down straps, etc.) may cause squeaks. Because of the nature of wood and construction methods, it is practically impossible to eliminate all squeaks during all seasons. Clearly, some squeaks are more objectionable than others.

3–3–2   **Observation:** A wood subfloor is uneven.

Performance Guideline: Subfloors shall not have more than a 1/4-inch ridge or depression within any 32-inch measurement. Measurements should not be made at imperfections that are characteristic of the code-approved material used. This guideline does not cover transition points between different materials.

Corrective Measure: The contractor will correct or repair the subfloor to meet the performance guideline.

3–3–3   **Observation:** A wood floor is out of square.

15

Performance Guideline: The diagonal of a triangle with sides of 12 feet and 16 feet along the edges of the floor shall be no more than 1/2-inch more nor less than 20 feet. *Remodeling Specific: The consumer and the contractor may agree to build a wood floor out of square in order to match or otherwise compensate for pre-existing conditions.*

Corrective Measure: The contractor will make the necessary modifications to any floor not complying with the performance guideline for squareness. The modification will produce a satisfactory appearance and may be either structural or cosmetic.

Discussion: Squareness is primarily an aesthetic consideration. Regularly repeated geometric patterns in floor and ceiling coverings show a gradually increasing or decreasing pattern along an out-of-square wall. The guideline tolerance of plus or minus 1/2-inch in the diagonal allows a maximum increasing or decreasing portion of about 3/8-inch in a 12-foot wall of a 12x16-foot room. However, a contractor and consumer may agree to build an addition out of square in order to keep a new exterior wall in line with an existing wall of an out-of-square house.

The corrective measure permits the contractor to make the modification in the most practical manner as long as "a satisfactory appearance" results.

3–3–4   **Observation:** A wood floor is out of level.

Performance Guideline: The floor should not slope more than 1/2-inch in 20 feet. Crowns and other lumber characteristics that meet the standards of the applicable grading organization for the grade and species used are not defects. Deflections due to overloading by the consumer are not the contractor's responsibility. *Remodeling Specific: The contractor and the consumer may agree to build an addition out of level in order to keep the floor of an addition on the same plane, and the roof ridge on the same line, as those of an existing, out-of-level structure, or to compensate for some other pre-existing condition.*

Corrective Measure: The contractor will make a reasonable and cost-effective effort to modify the floor that does not comply with the performance guideline.

Discussion: Sloped floors have both an aesthetic and functional consideration. Measurements for slope should be made across the room, not in a small area.

3–3–5   **Observation:** Deflection is observed in a floor system constructed of wood I-joists, floor trusses, or similar products.

Performance Guideline: All wood I-joists and other manufactured structural components in the floor system shall be sized and installed as provided in the manufacturers' instructions and code requirements.

Corrective Measure: The contractor will reinforce or modify as necessary any floor component not meeting the performance guideline.

Discussion: Deflection may indicate an aesthetic consideration independent of the strength and safety requirements of the product. When a consumer's preference is made known before construction, a higher standard may be agreed upon in writing by the contractor and the consumer.

3–3–6   **Observation:** *Remodeling Specific: Wood flooring is not level at the transition of an existing floor to a room addition floor.*

Performance Guideline: Flooring at a transition area shall not slope more than 1/8-inch over 6 inches unless a threshold is added. Overall step-down, unless previously agreed upon with the consumer, shall not exceed 1 1/8 inches. Variations caused by seasonal or temperature changes are not a defect.

Corrective Measure: The flooring transition shall be corrected to meet the performance guideline.

Discussion: All wood members shrink and expand seasonally, with variations in temperature and humidity, and with aging. After installation, 2x dimensional lumber can shrink up to 1/2-inch. If the flooring, subfloor, or underlayment was not purposely overlapped onto the existing floor, the resulting irregularity is not a defect, but a natural result and characteristic of the wood's aging process. The drier the house becomes, the more shrinkage may be experienced. Either the old or the new floors may slope along the floor joist span. Joists in older homes may have deflected under load. This and other conditions may cause a hump at the juncture of the old to new. If old and new flooring joists meet perpendicularly to each other, the first new floor joist running parallel to the old outside wall can fall that 1/2-inch out to the first parallel joist (14 1/2 inches into the new floor).

3–3–7   **Observation:** *Remodeling Specific: The floor pitches to one side in the door opening between the existing construction and the addition.*

17

Performance Guideline: If the pitch is the result of the floor of the existing dwelling not being level, then in most situations a transition threshold may be the most appropriate and acceptable means of addressing the condition.

Corrective Measure: The contractor will make a reasonable and cost-effective effort to meet the performance guidelines.

4

Walls

Wall Framing

4-1-1 **Observation:** A framed wall is not plumb.

Performance Guideline: The interior face of wood-framed walls shall not be more than 3/8-inch out of plumb for any 32 inches in any vertical measurement. *Remodeling Specific: The consumer and contractor may agree to intentionally build walls out of plumb to match the existing structure to accommodate or compensate for inaccuracies in the existing structure, and to disregard the performance guideline to match a pre-existing structural condition of the existing structure.*

Corrective Measure: The contractor will repair the wall to meet the performance guideline.

4-1-2 **Observation:** The wall is bowed.

Performance Guideline: Walls shall not bow more than 1/2-inch out of line within any 32-inch horizontal measurement, or 1/2-inch out of line within any 8-foot vertical measurement. *Remodeling Specific: If new wall cladding is installed on existing framed walls, the consumer and contractor may agree to straighten the wall as part of scope of work, to install new cladding over existing framing, and to disregard the performance guideline to match a pre-existing structural condition of the existing structure.*

Corrective Measure: The contractor will repair the wall to meet the performance guideline.

Discussion: All interior and exterior walls have slight variances in their finished surface. On occasion, the underlying framing may warp, twist, or bow after installation.

4-1-3 **Observation:** An exterior wall leaks because of improper caulking installation or failure of the caulking material.

Performance Guideline: Joints and cracks in exterior wall surfaces and around openings shall be caulked to prevent the entry of water.

Corrective Measure: One time only, the contractor will repair or caulk joints and cracks in exterior wall surfaces as required to correct deficiencies.

Discussion: Even when properly installed, caulking eventually will shrink and crack. Maintenance of caulking is the consumer's responsibility.

Wall Insulation

4–2–1   **Observation:** Wall insulation is insufficient.

Performance Guideline: The contractor shall install insulation according to R-values designated in the contract documents or local code, as applicable. Insulation shall be installed according to locally accepted practices.

Corrective Measure: The contractor will install insulation to meet the performance guideline.

Windows

4–3–1   **Observation:** A window is difficult to open or close.

Performance Guideline: Windows should require no greater operating force than that described in the manufacturer's instructions. *Remodeling Specific: The contractor is not responsible for inoperable windows not covered by the remodeling contract.*

Corrective Measure: The contractor will correct or repair the window as required to meet the performance guideline.

4–3–2   **Observation:** Window glass is broken and/or a screen is missing or damaged.

Performance Guideline: Glass should not be broken and screens should not be damaged at the time of substantial competition of the project. Screens required by the contract shall be installed.

Corrective Measure: Broken glass and/or missing or damaged screens reported to the contractor before closing will be installed or replaced. Broken glass and/or screens not reported prior to substantial completion of the project are the consumer's responsibility.

4–3–3   **Observation:** Mirror or glass surfaces are scratched.

Performance Guideline: Glass or mirror surfaces shall not have scratches visible from 10 feet under normal lighting conditions at the time of substantial completion of the project. *Remodeling Specific: This guideline does not apply to existing windows unless they are part of the remodeling contract or are damaged by the contractor. The consumer and contractor should examine existing windows prior to contract execution.*

Corrective Measure: The contractor shall replace any scratched glass or mirror surface if noted prior to substantial completion of the project.

4–3–4   **Observation:** During rains, water is observed on the interior corner of a glazed window unit.

Performance Guideline: Water leakage from improper installation is considered excessive. Leakage due to the manufacturer's design specifications is acceptable.

Corrective Measure: The contractor shall repair any deficiencies attributable to improper installation.

Discussion: Leakage at the glazing interface is covered under the manufacturer's warranty.

4–3–5   **Observation:** Window grids (muntins) fall or become out of level.

Performance Guideline: Window grids shall not disconnect, fall, or become out of level.

Corrective Measure: Window grids will be repaired or replaced at the contractor's discretion one time only.

4–3–6   **Observation:** A mirror backing is deteriorating.

Performance Guideline: While looking at the mirror, there should be no noticeable imperfections in the mirror as a result of damage to the mirror backing at the time of substantial completion of the project.

Corrective Measure: The contractor will replace or repair the mirror.

Exterior Doors

4–4–1 **Observation:** An exterior door is warped.

Performance Guideline: Exterior doors shall not warp to the extent that they become inoperable or cease to be weather-resistant. A 1/4-inch tolerance as measured diagonally from corner to corner is acceptable.

Corrective Measure: The contractor will correct or replace exterior doors that do not meet the performance guideline.

Discussion: Most exterior doors will warp to some degree due to the difference in the temperature and humidity between inside and outside surfaces; 1/4-inch across the plane of the door measured diagonally from corner to corner is an acceptable tolerance. Warping may also be caused by improper or incomplete finishing of the door including sides, top, and bottom. The contractor is not responsible for warpage if painting of doors is not within the contractor's scope of work.

4–4–2 **Observation:** Raw wood shows at the edges of an inset panel inserted into a wood exterior door during the manufacturing process.

Performance Guideline: This is a common occurrence in wood doors with panels.

Corrective Measure: Since this occurrence is common, no correction is required.

Discussion: Wood products expand and contract with changes in temperature and humidity. Wooden inserts are often loosely fitted into the rails to allow the inserts to move; this minimizes splitting of the panel or other damage to the door. The consumer is responsible for controlling temperature and humidity in the home to minimize these occurrences.

4–4–3 **Observation:** A wooden door panel is split.

Performance Guideline: A split in a panel shall not allow light to be visible through the door.

Corrective Measure: One time only, the contractor will repair, paint, or stain the split panel that does not meet the performance guideline. Caulking and fillers are acceptable. The repainted area may not match the remainder of the door or other doors on the house.

Discussion: Wooden inserts are loosely fitted into the door to allow the inserts to move; this minimizes splitting of the panel or other damage to the door. On occasion, a panel may become "locked" by paint or expansion of the edges with changes in temperature and humidity and no longer "float" between the rails. This may result in the panel splitting.

4–4–4   **Observation:** An exterior door sticks.

Performance Guideline: Exterior doors shall operate smoothly, except that doors may stick during occasional periods of high humidity or with variations in temperature.

Corrective Measure: The contractor will adjust or replace the door to meet the performance guideline.

Discussion: Exterior doors may warp or bind to some degree because of the difference in the temperature and/or humidity between inside and outside surfaces. The contractor is not responsible for warpage if painting of doors is not within the contractor's scope of work.

4–4–5   **Observation:** An exterior door will not shut completely.

Performance Guideline: Exterior doors shall shut completely.

Corrective Measure: The contractor will adjust or replace the door to meet the performance guideline.

Discussion: Exterior doors may warp or bind to some degree because of the difference in the temperature and/or humidity between inside and outside surfaces. The contractor is not responsible for warpage if painting of doors is not within the contractor's scope of work.

4–4–6   **Observation:** The plastic molding on the primary door behind the storm door melts from exposure to sunlight.

Performance Guideline: The plastic moldings behind storm doors should not melt if the storm panel is removed and reinstalled by the consumer as a part of normal seasonal maintenance operations (i.e., removed in the spring and reinstalled in the fall).

Corrective Measure: No corrective action is required.

Discussion: Plastic moldings may melt or deform if the exterior door is covered by a storm door panel during a warm season, or if it faces the sun. This is not a defect of the door, but a problem caused by the trapping of heat between the storm panel and the door. The consumer is also cautioned to follow the manufacturer's recommendations on painting the moldings with a dark color, with or without the use of a storm panel. Dark colors should be avoided.

4–4–7   **Observation:** Caulking or glazing on the primary door behind the storm door cracks or peels.

Performance Guideline: Glazing or caulking behind storm doors should not crack or peel if the storm panel is removed and installed by consumer as part of seasonal maintenance operations (i.e., removed in the spring and reinstalled in the fall).

Corrective Measure: No corrective measure is required.

Discussion: High temperatures may cause glazing and caulking to harden and/or fail prematurely if the door is covered by a storm panel during a warm season or if it faces the sun. This is not a defect of the door, caulking, or glazing, but a problem caused by the trapping of heat between the door and the storm panel. The consumer is reminded that dark colors tend to accumulate heat and are more likely to cause problems.

4–4–8   **Observation:** A door swings open or closed by the force of gravity.

Performance Guideline: Exterior doors shall not swing open or closed by the force of gravity alone. *Remodeling Specific: For remodeling projects, this guideline does not apply where a new door is installed in an existing wall that is out of plumb.*

Corrective Measure: The contractor will adjust the door to prevent it from swinging open or closed by the force of gravity.

4–4–9   **Observation:** Gaps are visible around an exterior door edge, doorjamb, and/or threshold.

Performance Guideline: Gaps between adjacent components shall not vary by more than 3/16-inch. *Remodeling Specific: This applies unless the existing building is out of square or plumb.*

Corrective Measure: The contractor will repair existing unit to meet performance guideline.

Discussion: Doors must have gaps at their perimeter to accommodate expansion/contraction due to variations in temperature and/or humidity and to enable the door to operate over a wide range of environmental conditions.

4-4-10 **Observation:** Exterior door hardware or kickplate has tarnished.

Performance Guideline: Finishes on door hardware or kickplates installed by the contractor are covered by the manufacturer's warranty.

Corrective Measure: The consumer should contact the manufacturer.

4-4-11 **Observation:** A sliding patio door or screen will not stay on track.

Performance Guideline: Sliding patio doors and screens shall slide properly on their tracks at the time of substantial completion of the project. The cleaning and maintenance necessary to preserve proper operation are consumer responsibilities.

Corrective Measure: The contractor shall repair the door or screen one time only.

Discussion: Proper operation should be verified by the consumer and the contractor at the time of substantial completion of the project.

4-4-12 **Observation:** A sliding patio door does not roll smoothly.

Performance Guideline: Sliding patio doors shall roll smoothly at the time of substantial completion of the project. The cleaning and maintenance necessary to preserve proper operation are consumer responsibilities.

Corrective Measure: The contractor shall repair the door one time only.

Discussion: Proper operation should be verified by the consumer and the contractor at the time of substantial completion of the project.

4-4-13 **Observation:** A doorknob, deadbolt, or lockset does not operate smoothly.

Performance Guideline: A doorknob, deadbolt, or lockset should not stick or bind during operation.

Corrective Measure: One time only, the contractor will adjust, repair, or replace knobs that are not damaged by abuse.

Exterior Finish

Wood and Hardboard Siding

4–5–1   **Observation:** Siding is bowed.

Performance Guideline: Bows exceeding 1/2-inch in 32 inches are considered excessive. *Remodeling Specific: If new wall covering is installed on existing framed walls, the consumer and contractor may agree to straighten out the walls as part of the scope of work. Alternatively, the parties may agree to install new wall covering over existing framing and disregard the performance guideline to match a pre-existing structural condition of the existing structure.*

Corrective Measure: The contractor will replace any wood lap siding with bows that does not meet the performance guideline, and will finish the replacement siding to match the existing siding as closely as practical.

Discussion: If the siding is fastened by nails driven into studs, expansion caused by changing relative temperatures and/or humidity may cause bulges or waves. Even with proper installation, siding will tend to bow inward and outward in adjacent stud spaces.

4–5–2   **Observation:** An edge or gap is visible between adjacent pieces of siding or siding panels and other materials.

Performance Guideline: Gaps wider than 3/16-inch are considered excessive. This guideline does not apply to adjacent pieces or panels that have shiplap or similar joints.

Corrective Measure: The contractor will repair gaps that do not meet the performance guideline.

Discussion: Proper repair can be effected by providing joint covers or by caulking the gap. This is important if the gaps were intentionally made for expansion joints. If the siding is painted, the contractor will paint the new caulking to match the existing caulking as closely as practical, but an exact match cannot be ensured.

4–5–3   **Observation:** Lap siding is not parallel with the course above or below.

Performance Guideline: A piece of lap siding may not be more than 1/2-inch off parallel with contiguous courses in any 20-foot measurement, unless the consumer and the contractor have previously agreed to disregard the performance guideline to match a pre-existing condition. *Remodeling Specific: The consumer and contractor may agree to install siding to match existing conditions on existing structure and to disregard the performance guideline for this item.*

Corrective Measure: The contractor will reinstall siding to meet the performance guideline for straightness, and will replace with new siding any siding damaged during removal.

Discussion: For remodeling projects, if the contractor and the consumer have agreed that the floor of an addition is to be on a different plane from an existing floor (e.g., out of level), the siding on the addition may not be parallel and in line with the existing siding.

4–5–4   **Observation:** Face nails are driven below the surface of the hardboard siding.

Performance Guideline: Siding nails should not be driven below the surface of hardboard siding such that visible fiber of the siding is exposed.

Corrective Measure: The contractor shall repair as necessary to meet performance guideline. The following repairs are appropriate in most instances: If visible fiber of hardboard siding is exposed, paint surface to coat fiber. If nail is 1/16 to 1/8-inch below the surface, fill or caulk and touch-up paint. If nail is more than 1/8-inch below the surface, fill or caulk and add an additional nail flush to the surface.

4–5–5   **Observation:** Siding boards have buckled.

Performance Guideline: Boards that project more than 3/16-inch from the face of adjacent boards are considered excessive.

Corrective Measure: The contractor will repair or replace any boards that don't meet the performance guideline.

Discussion: Buckling is caused by wood expanding as a result of increased temperature and/or relative humidity. It can be minimized by leaving space between the tongues and grooves to allow room for expansion and by storing the product outside for a few days to allow it to adjust to the ambient conditions prior to installation.

4–5–6   **Observation:** Cedar shakes or shingles have "bled" through paint or stain applied by the contractor.

Performance Guideline: Resins and extractives bleeding through paint or stain, or blackening of shakes or shingles is considered excessive. This performance guideline does not apply if "natural weathering" or semi-transparent stain is specified for the project.

Corrective Measure: One time only, the contractor will clean and treat shakes to provide a reasonable appearance and prevent further bleeding.

4–5–7   **Observation:** Siding has delaminated.

Performance Guideline: Siding shall not delaminate.

Corrective Measure: The contractor will replace delaminated siding that is not covered under the manufacturer's warranty, unless the delamination was caused by the consumer's actions or negligence. The repaired area may not precisely match the original siding.

4–5–8   **Observation:** Joints between siding have separated.

Performance Guideline: Joint separations exceeding 3/16-inch are considered excessive.

Corrective Measure: The contractor will caulk or repair siding as necessary to fill the joint. The repaired area may not match the original siding precisely.

Discussion: Plywood siding, like all wood products, will expand and contract with changes in temperature and/or humidity.

4–5–9   **Observation:** Siding is bowed.

Performance Guideline: Some waviness in siding is to be expected because of bows in studs. Bows exceeding 1/2-inch in 32 inches are considered excessive.

Corrective Measure: The contractor will repair or replace the siding to meet the guideline.

Discussion: Additional nails or screws may be installed to remove the bow.

Aluminum or Vinyl Lap Siding

4–5–10   **Observation:** Aluminum or vinyl siding is bowed or wavy.

Performance Guideline: Some waviness in aluminum or vinyl lap siding is to be expected because of bows in studs. Waves or similar distortions in aluminum or vinyl lap siding are considered excessive if they exceed 1/2-inch in 32 inches.

Corrective Measure: The contractor will correct any waves or distortions to comply with the performance guideline by reinstalling or replacing siding as necessary.

Discussion: This problem can be caused by the siding being nailed too tightly to the house instead of loosely "hung" near the center of the nail slots, or by not allowing adequate room for the siding to expand. Siding fasteners should be installed in the center of the nail slot with a 1/32-inch spacing (thickness of a dime) between the siding and the fastener to allow for expansion and contraction.

4–5–11   **Observation:** Nail stains are visible on siding or ceiling boards.

Performance Guideline: Stains exceeding 1/2-inch from the nail and readily visible from a distance in excess of 20 feet are considered excessive.

Corrective Measure: The contractor can choose to remove stains that do not meet the performance guideline.

Discussion: Stains can be caused by oxidation of nails or leaching of extractives from the wood. Use of galvanized nails (even double hot-dipped) will not necessarily prevent staining.

4–5–12   **Observation:** Siding is faded.

Performance Guideline: Any color siding, when exposed to the ultra-violet rays of the sun, will fade. Fading cannot be prevented by the contractor. However, panels installed on the same wall and under the same conditions should fade at the same rate.

Corrective Measure: No corrective action is required of the contractor. The consumer should contact the siding manufacturer.

Discussion: Color warranties are provided by the siding manufacturer. The consumer should contact the manufacturer with questions or claims regarding changes in color of vinyl or aluminum siding. Color and fade imperfections beyond an expected degree may be covered by the manufacturer's warranty, except where siding is shaded differently from the rest of the wall, such as under shutters or behind vegetation.

4–5–13 **Observation:** Aluminum or vinyl lap siding trim is loose.

Performance Guideline: Trim shall not separate from the house by more than 1/4-inch.

Corrective Measure: The contractor will reinstall trim as necessary to comply with the performance guideline.

Discussion: Vinyl siding and accessories should not be caulked in most circumstances, as it could impact the product's contraction and expansion characteristics.

4–5–14 **Observation:** Aluminum or vinyl lap siding courses are not parallel with eaves or wall openings.

Performance Guideline: Any piece of aluminum or vinyl lap siding more than 1/2-inch off parallel in 20 feet with a break such as an eave or wall opening is considered excessive. *Remodeling Specific: The consumer and contractor may agree to install siding to match existing conditions on the existing structure and to disregard the performance guideline for this item.*

Corrective Measure: The contractor will reinstall siding to comply with the performance guideline and will replace with new siding any siding damaged during removal.

Discussion: For remodeling projects, if the contractor and the owner agree that the floor of an addition is to be on a different plane from the existing floor (for example, a pre-existing out-of-level condition), the siding on the addition may not be parallel and in line with existing siding. Incorrect or inconsistent siding fastening can contribute to unparallel issues.

4–5–15 **Observation:** Nail heads show in aluminum or vinyl lap siding.

Performance Guideline: No nail heads in the field of the siding shall be exposed.

Corrective Measure: The contractor will install trim as necessary to cover the nails. Contractor will install proper trim accessories to avoid face nailing.

Discussion: Vinyl siding generally should not be face nailed. However, there are appropriate and typical occasions when a single face nail may be needed to reinforce a joint or fasten the siding to the wall when it is cut to fit around window frames, doors, roofs, or other obstructions on the wall.

In most cases (the only exception would be the top piece on a gable end), vinyl siding should never need to be face nailed when proper accessory products are used. For example, under a window application the trim (J-channel) can be utilized in conjunction with utility trim and snap-punching the top of the modified vinyl siding. If face nailing is the only option, a 1/8-inch diameter hole should be pre-drilled to allow for expansion and contraction.

4–5–16 **Observation:** An aluminum or vinyl lap siding trim accessory is loose from caulking at windows or other wall openings.

Performance Guideline: Siding trim accessories shall not separate from caulking at windows or other wall openings during the warranty period.

Corrective Measure: The contractor will repair or recaulk as necessary to eliminate the separation.

4–5–17 **Observation:** Aluminum or vinyl lap siding is cut crookedly.

Performance Guideline: Gaps shall comply with the manufacturer's guidelines unless the existing building is out of square or plumb. Cut edges of vinyl siding should always be covered by trim or receiving channels and should not be visible. Cuts should be made so that when properly installed in trim, edges are not visible. *Remodeling Specific: The consumer and contractor may agree to install siding to match conditions on the existing structure and to disregard the performance guideline for this item.*

Corrective Measure: The contractor will ensure that the appropriate trim/accessory is installed to eliminate potentially revealing site cuts. If cuts in siding panels are so uneven that they are not concealed by trim, the panel shall be replaced.

Discussion: Cut edges of vinyl siding should never be visible when proper trim and accessories are used.

4–5–18   **Observation:** Aluminum or vinyl lap siding is not correctly spaced from moldings.

Performance Guideline: Prescribed spacing between siding and accessory trim is typically 1/4-inch, or should comply with the manufacturer's installation instructions. *Remodeling Specific: The consumer and contractor may agree to install siding to match conditions on existing structure and to disregard the performance guideline for this item.*

Corrective Measure: The contractor will correct the spacing to meet the guideline.

Cement Board Siding

4–5–19   **Observation:** Cement board siding is cracked or chipped.

Performance Guideline: A cement product, this siding is susceptible to the same characteristic limitations as other cement products. Cracks more than 2 inches in length and 1/8-inch in width are considered excessive. Chips or dents not reported at time of substantial completion of the project are not covered.

Corrective Measure: Cracked or chipped cement board will be repaired or replaced as necessary, as determined by the contractor.

4–5–20   **Observation:** Cement board siding is improperly fastened.

Performance Guideline: Siding shall be nailed flush and perpendicular per the manufacturer's instructions. Staples shall not be used.

Corrective Measure: Overdriven nail heads or nails driven at an angle shall be filled with cementitious patching compound to match the existing area as closely as possible.

Discussion: The manufacturer's instructions include guidelines to reduce chipping or cracking of siding.

Masonry and Veneer

4–5–21   **Observation:** A masonry or veneer wall is cracked.

Performance Guideline: Cracks visible from distances in excess of 20 feet or larger than 1/4-inch in width are not acceptable.

Corrective Measure: The contractor will repair cracks in excess of the performance guideline by tuck pointing, patching, or painting. The contractor will not be responsible for color variation between the original and new mortar.

Discussion: Hairline cracks resulting from shrinkage and cracks due to minor settlement are common in masonry or veneer and do not necessarily represent a defect.

4–5–22   **Observation:** Cut bricks below openings in masonry walls are of different thickness.

Performance Guideline: Cut bricks used in the course directly below an opening shall not vary from one another in thickness by more than 1/4-inch. The smallest dimension of a cut brick should be greater than 1 inch.

Corrective Measure: The contractor will repair the wall to meet the performance guideline.

Discussion: Bricks are cut to achieve required dimensions at openings and ends of walls when it is not possible to match unit/mortar coursing.

4–5–23   **Observation:** A masonry or brick veneer course is not straight.

Performance Guideline: No point along the bottom of any course shall be more than 1/4-inch higher or lower than any other point within 10 feet along the bottom of the same course, or 1/2-inch in any length. *Remodeling Specific: The consumer and contractor may agree to install brick veneer to match conditions on the existing structure and to disregard the performance guideline for this item.*

Corrective Measure: The contractor will rebuild the wall as necessary to meet the performance guideline.

Discussion: Dimensional variations of the courses depend upon the variations in the brick selected.

4–5–24   **Observation:** Brick veneer is spalling.

Performance Guideline: Spalling of newly manufactured brick should not occur and is considered excessive. Spalling of used brick is acceptable.

Corrective Measure: The contractor will repair or replace newly manufactured bricks that have spalled. An exact match of brick and mortar cannot be assured.

4–5–25   **Observation:** Mortar stains are observed on exterior brick or stone.

Performance Guideline: Exterior brick and stone shall be free from mortar stains detracting from the appearance of the finished wall when viewed from a distance of 20 feet.

Corrective Measure: The contractor will clean the mortar stains to meet the performance guideline.

4–5–26   **Observation:** Efflorescence is present on the surface of masonry or mortar.

Performance Guideline: This is a common condition caused by moisture reacting with the soluble salts in the mortar.

Corrective Measure: No corrective actions are required of the contractor.

Discussion: Efflorescence is evidenced by the presence of a white film on the surface of masonry or mortar. It is a particularly common occurrence where masonry or concrete are in contact with high moisture levels as may be found in basements.

Stucco and Parge

4–5–27   **Observation:** An exterior stucco wall surface is cracked.

Performance Guideline: Cracks in exterior stucco wall surfaces shall not exceed 1/8-inch in width.

Corrective Measure: One time only, the contractor will repair cracks exceeding 1/8-inch in width. Caulking and touch-up painting are acceptable. An exact color or texture match may not be unattainable.

Discussion: "Stucco" includes cementitious coatings and similar synthetically based finishes.

4–5–28   **Observation:** The colors of exterior stucco walls do not match.

Performance Guideline: The colors of new exterior stucco walls may not perfectly match the colors of old exterior stucco walls, nor is it expected that exact matches will be attained for the same material that is applied on different days or under differing environmental conditions (e.g., temperature, humidity, etc.).

Corrective Measure: No corrective measure is required. Because of the unique nature of stucco finishes, exact match of color may not be possible.

Discussion: Coloring of stucco is affected by a number of variables. It is impractical to achieve a color match between stucco coatings applied at different times.

4–5–29   **Observation:** The textures of exterior stucco wall finishes do not match.

Performance Guideline: *Remodeling Specific: The texture of new exterior stucco walls may not perfectly match the textures of old exterior stucco walls.*

Corrective Measure: No corrective measure is required. Because of the unique nature of stucco finishes, exact match of texture finish may not be possible.

Discussion: "Stucco" includes cementitious coatings and similar synthetically based finishes. Approved samples prior to installation can minimize misunderstandings about color and texture.

4–5–30   **Observation:** Coating has separated from the base on an exterior stucco wall.

Performance Guideline: The coating shall not separate from the base on an exterior stucco wall during the warranty period.

Corrective Measure: The contractor will repair areas where the coating has separated from the base.

Discussion: Coloring of stucco is affected by a number of variables. It is impractical to achieve a color match between stucco coatings applied at different times.

4-5-31 **Observation:** Lath is visible through stucco.

Performance Guideline: Lath should not be visible through stucco, nor should the lath protrude through any portion of the stucco surface.

Corrective Measure: The contractor will make necessary corrections so that lath is not visible. The finish colors may not match.

4-5-32 **Observation:** Rust marks are observed on the stucco finish coat.

Performance Guideline: Rust marks on the stucco surface are considered excessive if more than 5 marks measuring more than 1 inch long occur per 100 square feet.

Corrective Measure: The contractor may repair or replace affected subsurface components, or seal the rusted areas and recolor the wall.

4-5-33 **Observation:** There is water damage to interior walls as a result of a leak in the stucco wall system.

Performance Guideline: Stucco walls should be constructed and flashed to prevent water penetration to the interior of the structure under normal weather and water conditions. Damage to the stucco system caused by external factors out of the contractor's control that result in water penetration are not the contractor's responsibility.

Corrective Measure: If water penetration is the result of a system failure and doesn't result from external factors, the contractor will make necessary repairs to prevent water penetration through the stucco wall system.

Discussion: Water penetration resulting from external factors such as windblown moisture or sprinkler systems are not the contractor's responsibility.

Exterior Trim

4-6-1 **Observation:** Gaps show in exterior trim.

Performance Guideline: Joints between exterior trim elements, including siding and masonry, shall not result in joints opened wider than 1/4-inch. In all cases, the exterior trim shall perform its function of excluding the elements.

Corrective Measure: The contractor will repair open joints that do not meet the performance guideline. Caulking is acceptable.

4–6–2   **Observation:** Exterior trim board is split.

Performance Guideline: Splits wider than 1/8-inch are considered excessive.

Corrective Measure: The contractor will repair splits by filling with a durable filler. Touch-up painting may not match the surrounding area.

4–6–3   **Observation:** Exterior trim board is bowed or twisted.

Performance Guideline: Bows and twists exceeding 3/8-inch in 8 feet are considered excessive.

Corrective Measure: The contractor will repair defects that do not meet the performance guideline by refastening or replacing deformed boards. Touch-up painting may not match the surrounding area.

4–6–4   **Observation:** Exterior trim board is cupped.

Performance Guideline: Cups exceeding 3/16-inch in 5 1/2 inches are considered excessive.

Corrective Measure: The contractor will repair defects that do not meet the performance guideline by refastening or replacing deformed boards. Touch-up painting may not match the surrounding area.

Paint, Stain, and Varnish

4–7–1   **Observation:** Exterior painting, staining, or refinishing is required because of repair work.

Performance Guideline: Repairs required under these performance guidelines shall be finished to match the immediate surrounding areas as closely as practical.

Corrective Measure: The contractor will finish repaired areas as indicated.

Discussion: Touch-up painting, staining, or refinishing may not match the surrounding area.

4–7–2   **Observation:** Exterior paint or stain has peeled, flaked, or physically deteriorated.

Performance Guideline: Exterior paints and stains shall not fail during the paint manufacturer's warranty period.

Corrective Measure: If exterior paint or stain has peeled, developed an alligator pattern, or blistered, the contractor will properly prepare and refinish affected areas and match the color as closely as practical. Where deterioration of the finish affects more than 50 percent of the piece of trim or wall area, the contractor will refinish the entire wall.

4–7–3   **Observation:** Exterior paint or stain has faded.

Performance Guideline: Fading of exterior paints and stains is common. The degree of fading depends on environmental conditions.

Corrective Measure: Because fading is a common occurrence in paint and stains, no corrective action is required.

4–7–4   **Observation:** Varnish or lacquer finishes have deteriorated.

Performance Guideline: Clear finishes used on exterior surfaces may deteriorate rapidly. This is beyond the contractor's control.

Corrective Measure: Heat and sunlight can cause rapid deterioration of clear finishes. Maintenance is the consumer's responsibility. No corrective action is required of the contractor.

4–7–5   **Observation:** There is paint or stain overspray on surfaces not intended for paint or stain.

Performance Guideline: Paint or stain overspray on surfaces not intended for paint or stain that is visible at a distance of 6 feet under normal natural lighting conditions is not acceptable.

Corrective Measure: The contractor shall clean affected surfaces without damaging the surface.

4–7–6   **Observation:** Cabinet stain is uneven. Cabinet paint is not uniform or is mismatched.

Performance Guideline: Uneven stain color on wood cabinets is considered acceptable and is a result of the natural wood grain.

Painted cabinets should appear uniform under normal lighting conditions at a distance of 6 feet.

Corrective Measure: The contractor will stain or paint the area as required to meet the performance guideline.

5

Roofs

Roof Structure

5-1-1   **Observation:** The roof ridge beam has deflected.

Performance Guideline: Roof ridge beam deflection greater than 1 inch in 8 feet is considered excessive. *Remodeling Specific: If this is not in the scope of work, the guideline will be disregarded.*

Corrective Measure: The contractor shall repair affected ridge beams that do not meet the performance guideline.

5-1-2   **Observation:** A rafter or ceiling joist bows (up or down).

Performance Guideline: Bows greater than 1 inch in 8 feet are excessive. *Remodeling Specific: If this is not in the scope of work, the guideline will be disregarded.*

Corrective Measure: The contractor shall repair affected rafters or joists that bow in excess of the performance guideline.

Roof Sheathing

5-2-1   **Observation:** Roof sheathing is wavy or appears bowed.

Performance Guideline: Roof sheathing shall not bow more than 1/2-inch in 2 feet. *Remodeling Specific: If new sheathing is installed over existing rafters, the sheathing will follow the bows of the existing rafters. The consumer and contractor should agree on whether or not the rafters are to be straightened. If they are not to be straightened, the performance guideline for this item will be disregarded.*

Corrective Measure: The contractor will straighten bowed roof sheathing as necessary to meet the performance guideline.

Discussion: In rare instances, the contractor might have to install blocking between the framing members to straighten the sheathing.

Roof Vents

5–3–1   **Observation:** An attic vent or louver leaks.

Performance Guideline: Attic vents and louvers shall not leak. However, infiltration of wind-driven rain and snow are not considered leaks and are beyond the control of the contractor.

Corrective Measure: The contractor shall repair or replace the roof vents as necessary to meet the performance guideline.

Roof Installation and Leaks

Asphalt Shingles

5–4–1   **Observation:** The roof or flashing leaks.

Performance Guideline: Roofs and flashing shall not leak under normal conditions.

Corrective Measure: The contractor will repair any verified roof or flashing leaks not caused by ice build-up, leaves, debris, abnormal conditions, or the consumer's actions or negligence.

Discussion: It is the consumer's responsibility to keep the roof drains, gutters, and downspouts free of ice and debris.

5–4–2   **Observation:** Ice builds up on the roof.

Performance Guideline: During prolonged cold spells ice is likely to build up on a roof, especially at the eaves. This condition naturally can occur when snow and ice accumulates.

Corrective Measure: No action is required of the contractor. Prevention of ice build-up on the roof is a consumer maintenance item.

5–4–3   **Observation:** Shingles have blown off.

Performance Guideline: Shingles shall not blow off in winds less than the manufacturer's warranty statement or applicable building codes.

Corrective Measure: If shingles were not installed properly, they will be repaired or replaced in the affected area.

5-4-4   **Observation:** Shingles slide off the roof.

Performance Guideline: The contractor shall ensure that shingles are installed in accordance with the manufacturer's instructions.

Corrective Measure: The contractor shall evaluate and replace shingles that slide off the roof.

Discussion: Correctly installed shingles are covered by the manufacturer's warranty.

5-4-5   **Observation:** Shingles are not horizontally aligned.

Performance Guideline: Shingles should be installed according to the manufacturer's instructions. *Remodeling Specific: The consumer and the contractor may agree prior to installation that the horizontal line of shingles on the roof of an addition need not line up with those of the existing structure if the floors (and hence, the eaves and ridge) are not to be built on the same plane.*

Corrective Measure: The contractor will remove shingles that do not meet the performance guideline, and will repair or replace them with new shingles that are properly aligned.

Discussion: The bottom edge of dimensional shingles may be irregular; the irregularity is an inherent part of the design.

5-4-6   **Observation:** New shingles do not match existing shingles.

Performance Guideline: Because of weathering and manufacturing variations, the color of new shingles will not exactly match the color of existing shingles.

Corrective Measure: The contractor is not responsible for precisely matching the color of existing shingles.

5-4-7   **Observation:** Asphalt shingle edges or corners are curled or cupped.

Performance Guideline: Asphalt shingle edges and corners shall not curl or cup more than 1/2-inch.

Corrective Measure: No corrective action is required of the contractor. Cupping in excess of 1/2-inch should be reported to the manufacturer.

5-4-8   **Observation:** Asphalt shingles do not overhang the edges of the roof, or hang too far over the edges of the roof.

Performance Guideline: Asphalt shingles shall overhang roof edges by not less than 1/4-inch, and not more than 3/4-inch unless the manufacturer's instructions indicate otherwise.

Corrective Measure: The contractor will reposition or replace shingles as necessary to meet the performance guideline.

5-4-9   **Observation:** Shading or a shadowing pattern is observed on a new shingle roof.

Performance Guideline: Shading or shadowing is a defect only if it results from failure to use shingles of the type specified in the contract.

Corrective Measure: The contractor will replace shingles not conforming to the contractual requirements.

5-4-10   **Observation:** Asphalt shingles have developed surface buckling.

Performance Guideline: Asphalt shingle surfaces need not be perfectly flat. Buckling higher than 1/4-inch is considered excessive.

Corrective Measure: The contractor will repair or replace the affected shingles to meet the performance guideline.

5-4-11   **Observation:** Sheathing nails have loosened from framing and raised asphalt shingles.

Performance Guideline: Nails shall not loosen from roof sheathing to raise asphalt shingles from surface.

Corrective Measure: The contractor shall repair all areas as necessary to meet the performance guideline.

Discussion: It is not uncommon for nails to "work themselves out" due to variations in temperature. The contractor can re-drive or remove and replace fasteners that withdraw from the framing. Any resulting holes should be sealed or the shingle should be replaced (a perfect color/shade match cannot be assured).

5-4-12   **Observation:** Roofing nails are exposed at the ridge or hip of a roof.

Performance Guideline: Nail heads shall be sealed.

Corrective Measure: The contractor shall repair areas to meet the performance guideline.

5–4–13   **Observation:** Holes from construction activities are found in asphalt shingles.

Performance Guideline: Holes from construction activities shall be flashed or sealed below the asphalt shingle tab to prevent leakage. If the patch is visible from the ground, the shingle should be replaced.

Corrective Measure: The contractor will repair or replace the affected shingles to meet the performance guideline.

5–4–14   **Observation:** *Remodeling Specific: Existing roof shingles are telegraphing through new asphalt shingles.*

Performance Guideline: *Remodeling Specific: Some telegraphing is common when re-roofing over existing roofing.*

Corrective Measure: Because this is a common occurrence, no corrective action is required.

Roll Roofing

5–4–15   **Observation:** Water is trapped under roll roofing.

Performance Guideline: Water shall not become trapped under roll roofing.

Corrective Measure: If water becomes trapped under roll roofing during the warranty period, the contractor will repair or replace the roofing as necessary to meet the performance guideline.

5–4–16   **Observation:** Roofing is blistered but does not leak.

Performance Guideline: Surface blistering of roll roofing is caused by unusual conditions of heat and humidity acting on the asphalt and is a common occurrence.

Corrective Measure: Because this is a common occurrence, no action is required.

5–4–17   **Observation:** Water is standing on a flat roof.

Performance Guideline: Water shall drain from a flat roof except for minor ponding within 24 hours of a rainfall. Minor ponding shall not exceed 3/8-inch in depth.

Corrective Measure: The contractor will take corrective action to ensure proper drainage of the roof.

Chimney

5–5–1   **Observation:** A crack in a masonry chimney cap or crown causes leakage.

Performance Guideline: It is common for caps to crack due to expansion and contraction. As a result, leaks may occur.

Corrective Measure: If cracking causes leakage the contractor will repair the cap or crown. Caulking or other sealant is acceptable.

5–5–2   **Observation:** New chimney flashing leaks.

Performance Guideline: New chimney flashing shall not leak under normal conditions.

Corrective Measure: The contractor will repair leaks in new chimney flashing that are not caused by ice build-up, other common occurrences, or by the consumer's actions or negligence.

Discussion: The accumulation of ice and snow on the roof is a natural occurrence and cannot be prevented by the contractor.

Gutters and Downspouts

5–6–1   **Observation:** The gutter or downspout leaks.

Performance Guideline: Gutters and downspouts shall not leak.

Corrective Measure: The contractor will repair leaks in gutters and downspouts. Sealants are acceptable.

5–6–2   **Observation:** The gutter overflows during a heavy rain.

Performance Guideline: Gutters may overflow during a heavy rain.

Corrective Measure: The contractor shall repair the gutter if it overflows during normal rains.

Discussion: The consumer is responsible for keeping gutters and downspouts free from debris that could cause overflow.

5–6–3   **Observation:** Water remains in the gutter after a rain.

Performance Guideline: The water level shall not exceed 1/2-inch in depth if the gutter is unobstructed by ice, snow, or debris.

Corrective Measure: The contractor will repair the gutter to meet the performance guideline. The consumer is responsible for maintaining gutters and downspouts and keeping them unobstructed.

Discussion: Contractors usually install residential gutters with minimal slope in order to maintain an attractive appearance. Installing gutters with 1/32-inch drop in 1 foot generally will prevent water from standing in the gutters. Even so, small amounts of water may remain in some sections of the gutter for a time after a rain. In areas with heavy rainfall and/or ice build-up, a steeper pitch or additional downspouts may be desirable.

Skylights

5–7–1   **Observation:** A skylight leaks.

Performance Guideline: Skylights shall be installed in accordance with the manufacturer's instructions. Leaks resulting from improper installation are considered excessive. Condensation on interior surfaces is not a leak and is not considered a defect.

Corrective Measure: The contractor will repair any improperly installed skylight to meet the performance guideline.

Discussion: Condensation on interior surfaces is not a leak.

Plumbing

Note: *Remodeling Specific: The contractor is responsible only for areas of the property worked on and specified in the contract, and not for the entire house.*

Water Supply System

6–1–1   **Observation:** A pipe or fitting leaks.

Performance Guideline: No leaks of any kind shall exist in any water pipe or fitting. *Remodeling Specific: See Note at beginning of chapter.*

Corrective Measure: The contractor will make repairs to eliminate leakage.

6–1–2   **Observation:** Condensation is observed on pipes, fixtures, and plumbing supply lines.

Performance Guideline: Condensation on pipes, fixtures, and plumbing supply lines may occur at certain combinations of temperature and indoor humidity. *Remodeling Specific: See Note at beginning of chapter.*

Corrective Measure: The consumer is responsible for controlling humidity in the home.

Discussion: The consumer may insulate pipes and supply lines.

6–1–3   **Observation:** A faucet or valve leaks.

Performance Guideline: No faucet or valve shall leak as a result of defects in material or workmanship. *Remodeling Specific: See Note at beginning of chapter.*

Corrective Measure: The contractor will repair or replace the leaking faucet or valve.

6–1–4   **Observation:** Water in a plumbing pipe freezes, and the pipe bursts.

Performance Guideline: Drain, waste, vent, and water pipes shall be adequately protected to reduce the possibility of freezing at the design temperatures and based on the applicable building or plumbing code. *Remodeling Specific: See Note at beginning of chapter.*

Corrective Measure: The contractor will correct situations not meeting the applicable code. The consumer is responsible for draining or otherwise protecting pipes and exterior faucets exposed to freezing temperatures.

6–1–5   **Observation:** The water supply system fails to deliver water.

Performance Guideline: All on-site service connections to the municipal water main or private water supply are the responsibility of the contractor. *Remodeling Specific: See Note at beginning of chapter.*

Corrective Measure: The contractor will repair the water supply system if the failure results from improper installation or failure of materials and if the connections are a part of the construction agreement. Conditions beyond the control of the contractor that disrupt or eliminate the water supply are not covered.

6–1–6   **Observation:** A water pipe is noisy.

Performance Guideline: Because of the flow of water and pipe expansion/contraction, the water piping system will emit some noise. However, the pipes should not make the pounding noise called "water hammer." *Remodeling Specific: See Note at beginning of chapter.*

Corrective Measure: The contractor cannot eliminate all noises caused by water flow and pipe expansion/contraction. However, the contractor will provide the "water hammer" protection required by the applicable plumbing code.

Plumbing Fixtures

6–2–1   **Observation:** The bathtub or shower leaks.

Performance Guideline: Bathtubs and showers shall not leak. *Remodeling Specific: See Note at beginning of chapter.*

Corrective Measure: The contractor will repair bathtub or shower leaks as necessary to meet the performance guideline.

Discussion: Proper repair can be effected by sealing areas around tubs and showers. The consumer is responsible for maintaining caulk seals from occupancy onward.

6–2–2   **Observation:** A plumbing fixture, appliance, or trim fitting is defective.

Performance Guideline: Plumbing fixtures, appliances, and trim fittings shall not be damaged at the time of substantial completion of the project. *Remodeling Specific: See Note at beginning of chapter.*

Corrective Measure: No action is required of the contractor. Defective trim fittings, appliances, and fixtures are covered under the manufacturer's warranty.

6–2–3   **Observation:** The surface of a plumbing fixture is cracked or chipped.

Performance Guideline: Cracks and chips in surfaces of bathtubs and sinks are considered excessive if they are visible from 3 feet in normal lighting conditions. *Remodeling Specific: See Note at beginning of chapter.*

Corrective Measure: The contractor is not responsible for repairs unless the damage is reported to the contractor prior to substantial completion of the project. If the problem is the result of a manufacturing defect, the manufacturer's warranty is in effect.

Discussion: Fiberglass and acrylic fixtures often can be repaired.

6–2–4   **Observation:** A fiberglass tub or shower enclosure base flexes.

Performance Guideline: The tub or showers are to be installed according to the manufacturer's instructions. *Remodeling Specific: See Note at beginning of chapter.*

Corrective Measure: The contractor shall repair the base to meet the performance guideline.

6–2–5   **Observation:** A vanity top is cracked.

Performance Guideline: Vanity tops shall not have cracks when installed with proper sealants. *Remodeling Specific: See Note at beginning of chapter.*

Corrective Measure: The contractor shall repair or replace the

vanity top to meet the performance guidelines. Cracks must be
noted prior to substantial completion of the project.

Sanitary Sewer or Septic System

6–3–1   **Observation:** A sewer, fixture, or drain is clogged.

Performance Guideline: Sewers, fixtures, and drains shall drain.
Remodeling Specific: See Note at beginning of chapter.

Corrective Measure: The contractor is not responsible for sewers,
fixtures, and drains that are clogged because of the consumer's
actions or negligence. If a problem occurs, the consumer should
consult the contractor for corrective action. If defective
installation is the cause, the contractor is responsible for
correcting the problem. If the consumer's actions or negligence is
the cause, the consumer is responsible for correcting the problem.

Discussion: With respect to septic systems, consumer actions
that constitute negligence under this guideline include but are not
limited to the following:

- Connection of sump pump, roof drains, or backwash from
  a water conditioner into the system.
- Placement of non-biodegradable items into the system.
- Use of a food waste disposer not supplied or approved by
  the contractor.
- Placement of surfaces not permeable to water over the
  disposal area of the system.
- Allowing vehicles to drive or park over the disposal area of
  the system.
- Failure to pump out the septic tank periodically, as
  required.
- Use that exceeds the system's design standards.
- Allowing water to pond over the disposal area.

Electrical

Note: *Remodeling Specific: The contractor is responsible only for areas of the property worked on and specified in the contract, and not for the entire house.*

Fuses and Circuit Breakers

7–1–1   **Observation:** A fuse blows or a circuit breaker trips.

Performance Guideline: Fuses and circuit breakers shall not be tripped by normal usage. *Remodeling Specific: See Note at beginning of chapter.*

Corrective Measure: The contractor will check wiring circuits and components for conformity with applicable electrical code requirements. The contractor will correct noncompliant elements.

Discussion: Blown fuses and tripped breakers are symptoms of a problem in some part of the electrical system in the home or some consumer product connected to the system. Although defective components are possible, most electrical malfunctions are caused by consumer-owned fixtures and appliances. The consumer should unplug or disconnect fixtures and appliances on the circuit and then replace the fuse or reset the breaker. If the problem recurs, the contractor should be notified.

7–1–2   **Observation:** A ground fault circuit interrupter (GFCI) or arc fault circuit interrupter (AFCI) trips frequently.

Performance Guideline: Ground fault and arc fault circuit interrupters shall perform as designed. *Remodeling Specific: See Note at beginning of chapter.*

Corrective Measure: The contractor will install ground fault and arc fault circuit interrupters in accordance with applicable electrical codes. Tripping is to be expected and is not covered unless it is caused by a component failure or incorrect installation.

Discussion: Both ground fault and arc fault circuit interrupters are very sensitive devices and are easily tripped. GFCIs protect outlets in wet areas (for example, bathrooms, kitchens, garages, exterior, etc.). Outlets protected by GFCIs may be connected in series; it may not be readily apparent that an inoperative convenience outlet is the result of a tripped GFCI in another room (and not necessarily in the electrical panel). AFCIs sometimes are installed to protect bedroom circuits. The most common cause of tripping by AFCIs is damaged cords or plugs on consumers' lamps, small appliances, or other devices. AFCIs are usually found in the electrical panel.

Outlets and Lights

7–2–1   **Observation:** Electrical outlets, switches, or fixtures malfunction.

Performance Guideline: All electrical outlets, switches, and fixtures shall operate as designed. *Remodeling Specific: See Note at beginning of chapter.*

Corrective Measure: The contractor will repair or replace malfunctioning electrical outlets, switches, and fixtures, if supplied and installed by the contractor.

7–2–2   **Observation:** Wiring fails to carry its designed load.

Performance Guideline: Wiring shall be capable of carrying the designed load for normal residential use. *Remodeling Specific: See Note at beginning of chapter.*

Corrective Measure: The contractor will verify that wiring conforms to applicable electrical code requirements. The contractor will repair wiring not conforming to code.

7–2–3   **Observation:** A light fixture is tarnished.

Performance Guideline: Finishes on light fixtures may be covered under the manufacturer's warranty. *Remodeling Specific: See Note at beginning of chapter.*

Corrective Measure: No action is required of the contractor. Consumer should contact manufacturer.

7–2–4   **Observation:** Receptacle or switch covers protrude from the wall.

Performance Guideline: Receptacle or switch covers should not be more than 1/16-inch from the adjoining wall surface. *Remodeling Specific: See Note at beginning of chapter.*

Corrective Measure: The contractor will adjust the covers to meet performance guideline.

7–2–5   **Observation:** The consumer's 220-volt appliance cord does not fit the outlet provided by the contractor.

Performance Guideline: The contractor shall install electrical outlets required by applicable electrical code. *Remodeling Specific: See Note at beginning of chapter.*

Corrective Measure: No action is required of the contractor.

Discussion: The consumer is responsible for obtaining an appliance cord that fits the outlets provided by the contractor.

Fans

7–3–1   **Observation:** A ceiling fan vibrates excessively and/or is noisy.

Performance Guideline: The contractor shall install ceiling fans in accordance with the manufacturer's instructions (including blade balances). *Remodeling Specific: See Note at beginning of chapter.*

Corrective Measure: The contractor shall correct any fan installation not in accordance with the performance guideline if the fan was supplied and installed by the contractor.

7–3–2   **Observation:** An exhaust fan discharges into attic or crawl space.

Performance Guideline: Fans shall discharge as required by applicable codes. *Remodeling Specific: See Note at beginning of chapter.*

Corrective Measure: The contractor shall repair to meet performance guideline.

Smoke Detectors

7–4–1   **Observation:** A smoke detector "chirps."

Performance Guideline: A smoke detector should not "chirp" at substantial completion of the project. *Remodeling Specific: See Note at beginning of chapter.*

Corrective Measure: The contractor will repair or replace the smoke detector to eliminate chirping.

Discussion: Most smoke detectors are powered by both the home's electrical power and a backup battery. "Chirping" is an indication that the battery is weak or is not installed. If the chirping occurs on a new smoke detector, the contractor will check the battery, verify that the detector is wired correctly, and replace the device if necessary. Safety officials recommend that consumers change the batteries in smoke detectors semi-annually when daylight-saving time begins and ends.

Interior Climate Control

Note: Remodeling Specific: The contractor is responsible only for areas of the property worked on and specified in the contract, and not for the entire house.

Air Infiltration and Drafts

8–1–1   **Observation:** Air infiltrates around exterior doors or windows.

Performance Guideline: Some infiltration is usually noticeable around doors and windows, especially during high winds. No daylight shall be visible around the frame when the window or door is closed. *Remodeling Specific: See Note at beginning of chapter.*

Corrective Measure: The contractor shall repair to meet the performance guideline.

Discussion: Proper repair can be performed by adjusting or installing weather stripping around doors and windows. In high-wind areas, the consumer may elect to have storm windows and doors installed to further reduce drafts.

8–1–2   **Observation:** A draft comes through an electrical outlet.

Performance Guideline: Electrical outlets and switch boxes on exterior walls may allow cold air to flow through or around an outlet into a room. *Remodeling Specific: See Note at beginning of chapter.*

Corrective Measure: No action is required of the contractor. The consumer may elect to install foam insulation pads under switch and outlet plates to help decrease drafts.

Humidity Control and Condensation

8–2–1   **Observation:** Water, ice, or frost is observed on a window.

Performance Guideline: Windows will be installed in accordance with the manufacturer's instructions and applicable building code. *Remodeling Specific: See Note at beginning of chapter.*

Corrective Measure: No action is required of the contractor unless the water, ice, or frost is directly attributed to faulty installation (i.e., that deviates from the manufacturer's instructions and/or applicable building code).

Discussion: Condensation usually results from conditions beyond the contractor's control. Moisture in the air can condense into water and collect on cold surfaces, particularly in the winter months when the outside temperature is low. Blinds and drapes can prevent air within the building envelope from moving across the cold surface and picking up the moisture. Occasional condensation (water) in the kitchen, bath, or laundry area is common. It is the consumer's responsibility to maintain proper humidity by properly operating heating and cooling systems and allowing moving air within the home to flow over the interior surface of the windows.

8–2–2   **Observation:** The ductwork makes noises.

Performance Guideline: Ductwork will be constructed and installed in accordance with applicable mechanical code requirements. *Remodeling Specific: See Note at beginning of chapter.*

Corrective Measure: Unless the duct is not in compliance with the local code, no corrective action is required.

Discussion: Metal expands when it is heated and contracts when it cools. The "ticking" or "crackling" sounds caused by the metal's movement are common.

8–2–3   **Observation:** The ductwork produces excessively loud noises commonly known as "oil canning."

Performance Guideline: The stiffening of the ductwork and the thickness of the metal used shall be such that ducts do not "oil can." The booming noise caused by oil canning is considered excessive. *Remodeling Specific: See Note at beginning of chapter.*

Corrective Measure: The contractor will correct the ductwork to eliminate noise caused by oil canning.

8–2–4   **Observation:** There is airflow noise at a register.

Performance Guideline: The register should be correctly installed according to the manufacturer's instructions. *Remodeling Specific: See Note at beginning of chapter.*

Corrective Measure: No action is required unless registers are not installed in accordance with manufacturer's instructions.

Discussion: Under certain conditions, some noise may be experienced with the normal flow of air, even when registers are installed correctly. See the manufacturer's instructions.

8–2–5   **Observation:** The air handler or furnace vibrates.

Performance Guideline These items shall be installed in accordance with the manufacturer's instructions and applicable codes. *Remodeling Specific: See Note at beginning of chapter.*

Corrective Measure: If installed incorrectly, the contractor will correct the items according to the manufacturer's instructions and code requirements.

Discussion: Under certain conditions some vibrating may be experienced with the normal flow of air, even when air handlers and furnaces are installed correctly. See the manufacturer's instructions.

8–2–6   **Observation:** The ductwork is separated or detached.

Performance Guideline: Ductwork shall remain intact and securely fastened. *Remodeling Specific: See Note at beginning of chapter.*

Corrective Measure: The contractor will reattach and secure all separated or unattached ductwork.

Heating System

8–3–1   **Observation:** The heating system is inadequate.

Performance Guideline: The heating system shall be capable of producing an inside temperature of 70 degrees Fahrenheit, as measured in the center of each room at a height of 5 feet above the floor under local, outdoor winter design conditions. National, state, or local energy codes shall supersede this performance guideline where such codes have been locally adopted. *Remodeling Specific: See Note at beginning of chapter.*

Corrective Measure: The contractor will correct the heating system to provide the required temperature in accordance with the performance guideline or applicable code requirements. However, the consumer will be responsible for balancing dampers and registers and for making other necessary minor adjustments.

Discussion: For new living spaces created by remodeling jobs, heating guidelines may not apply to areas where living space has been created without providing additional heating and/or resizing the ductwork.

Central Air-Conditioning System

8-4-1   **Observation:** The cooling of rooms is inadequate.

Performance Guideline: If air conditioning is installed by the contractor, the cooling system shall be capable of maintaining a temperature of 78 degrees Fahrenheit, as measured in the center of each room at a height of 5 feet above the floor under local outdoor summer design conditions. In the case of outside temperatures exceeding 95 degrees Fahrenheit, the system shall keep the inside temperature 15 degrees Fahrenheit cooler than the outside temperature. National, state, or local codes shall supersede this guideline where such codes have been locally adopted. *Remodeling Specific: See Note at beginning of chapter.*

Corrective Measure: The contractor will correct the cooling system to provide the required temperature in accordance with the applicable code requirements.

Discussion: For new living spaces created by remodeling jobs, cooling guidelines may not apply to areas where living space has been created without providing additional cooling and/or resizing the ductwork.

8-4-2   **Observation:** A condensate line is clogged.

Performance Guideline: Condensate lines must be free of all clogs to operate properly. *Remodeling Specific: See Note at beginning of chapter.*

Corrective Measure: Condensate lines will eventually clog under normal use. The contractor will provide unobstructed condensate lines at the time of substantial completion of the project. The consumer is responsible for maintaining them in that condition.

8-4-3   **Observation:** There is a refrigerant leak.

Performance Guideline: Refrigerant lines and fittings shall not leak during normal operation. *Remodeling Specific: See Note at beginning of chapter.*

Corrective Measure: The contractor will repair leaking refrigerant lines and recharge the air-conditioning unit unless the damage was caused by the consumer's actions or negligence.

8-4-4   **Observation:** There is condensation on the outside of air handlers and ducts.

Performance Guideline: Moisture may condense on the exterior surfaces of air handlers and ducts under some temperature differences and high humidity levels. *Remodeling Specific: See Note at beginning of chapter.*

Corrective Measure: No action is required of the contractor, unless the condensation is directly attributed to faulty installation.

Discussion: Condensation usually results from conditions beyond the contractor's control. Moisture in the air can condense (to form water) and collect on cold duct surfaces, particularly in the summer months when the outside humidity is high.

8-4-5   **Observation:** Kitchen or bath fans allow air infiltration.

Performance Guideline: Bath and kitchen fans shall be installed in accordance with the manufacturer's instructions and code requirements. *Remodeling Specific: See Note at beginning of chapter.*

Corrective Measure: No action is required of the contractor if fans meet the guideline.

Discussion: It is possible for outside air to enter the house through a ventilation fan. The dampers in most fans do not seal tightly. It is possible for the damper to be lodged open due to animal activity (including nesting in the outside opening), or the accumulation of grease, lint, and other debris. Maintenance of ventilating fans is the consumer's responsibility.

8-4-6   **Observation:** HVAC vent or register covers protrude more than 1/16-inch from a smooth wall or ceiling surface.

Performance Guideline: Registers shall not protrude more than 1/16-inch from the wall surface at the time of substantial completion of the project. *Remodeling Specific: See Note at beginning of chapter.*

Corrective Measure: The contractor shall comply with the performance guideline.

Discussion: Registers and grills may deflect over time. This can result in gaps occurring between the grill or register and the wall or ceiling. As long as the vent or register is securely attached, this is not a warranty item.

Interior

Note: *Remodeling Specific: The contractor is responsible only for areas of the property worked on and specified in the contract, and not for the entire house.*

Interior Doors

9–1–1   **Observation:** An interior door is warped.

Performance Guideline: Interior doors (full openings) shall not warp in excess of 1/4-inch. *Remodeling Specific: See Note at beginning of chapter.*

Corrective Measure: The contractor will correct or replace and refinish defective doors to match existing doors as nearly as practical.

Discussion: In bathroom or utility areas, exhaust fans or an open window must be used to minimize moisture to prevent warpage of door units. If the consumer is responsible for painting the door, the contractor is not responsible.

9–1–2   **Observation:** Bifold doors come off their tracks during normal operation.

Performance Guideline: Bifold doors shall slide properly on their tracks at the time of substantial completion of the project. Cleaning and maintenance necessary to preserve proper operation are consumer responsibilities. *Remodeling Specific: See Note at beginning of chapter.*

Corrective Measure: One time only, the contractor will repair any bifold door that will not stay on its track during normal operation.

Discussion: Proper operation should be verified by the consumer and the contractor at the time of substantial completion of the project.

9–1–3   **Observation:** A pocket door rubs in its pocket during normal operation.

Performance Guideline: Pocket doors shall not rub in their pockets during normal operation if they are installed according to the manufacturer's instructions. *Remodeling Specific: See Note at beginning of chapter.*

Corrective Measure: One time only, the contractor will repair the pocket door to meet the performance guideline.

Discussion: Pocket doors commonly rub, stick, or derail due to the inherent nature of the product. It is common, however, for the door to operate against the guides provided by the manufacturer.

9–1–4   **Observation:** A wooden door panel has shrunk or split.

Performance Guideline: Wooden door panels shall not split to the point that light is visible through the door. *Remodeling Specific: See Note at beginning of chapter.*

Corrective Measure: One time only, the contractor will fill splits in the door panel with wood filler and will match the paint or stain as closely as practical.

9–1–5   **Observation:** A door rubs on jambs or contractor-installed floor covering.

Performance Guideline: Doors shall operate smoothly. *Remodeling Specific: See Note at beginning of chapter.*

Corrective Measure: One time only, the contractor will repair the door as necessary to meet the performance guideline.

9–1–6   **Observation:** A door edge is not parallel to the door jamb.

Performance Guideline: When the contractor installs the door frame and door, the door edge shall be within 3/16-inch of parallel to the door jamb. *Remodeling Specific: Where the contractor installs the door in an existing frame that is out of square, the guideline does not apply. See Note at beginning of chapter.*

Corrective Measure: The contractor will adjust the door as necessary to meet the guideline one time.

9–1–7   **Observation:** A door swings open or closed by the force of gravity.

Performance Guideline: Doors shall not swing open or closed by the force of gravity alone. *Remodeling Specific: This guideline does not* *apply where a door is installed in an existing wall that is out of plumb. See Note at beginning of chapter.*

Corrective Measure: The contractor will adjust the door as necessary to meet the guideline one time.

9–1–8   **Observation:** Interior doors do not operate smoothly.

Performance Guideline: Doors shall move smoothly with limited resistance. *Remodeling Specific: See Note at beginning of chapter.*

Corrective Measure: The contractor shall repair door operation to meet the performance guideline one time.

9–1–9   **Observation:** A door knob or latch does not operate smoothly.

Performance Guideline: A door knob or latch should not stick or bind during operation.

Corrective Measure: The contractor will adjust, repair, or replace knobs or latches that are not damaged by abuse one time.

Interior Stairs

9–2–1   **Observation:** An interior stair tread deflects too much.

Performance Guideline: The maximum vertical deflection of an interior stair tread shall not exceed 1/8-inch at 200 pounds of force.

Corrective Measure: The contractor will repair the stair to meet the performance guideline.

9–2–2   **Observation:** Gaps exist between interior stair risers, treads, and/or skirts.

Performance Guideline: Gaps between adjoining parts that are designed to meet flush shall not exceed 1/8-inch in width.

Corrective Measure: The contractor will repair the gap with filler or will replace the parts as necessary to meet the performance guideline.

9–2–3   **Observation:** A stair riser or tread squeaks.

Performance Guideline: Loud squeaks caused by a loose stair riser or tread are considered excessive; however, totally squeak-proof stair risers or treads cannot be guaranteed.

Corrective Measure: The contractor will refasten any loose risers or treads or take other reasonable and cost-effective corrective action to eliminate squeaking without removing treads or ceiling finishes.

Discussion: Squeaks in risers or treads may occur when a riser has come loose from the tread, and is deflected by the weight of a person and rubs against the nails that hold it in place. Movement may occur between the riser and the tread or other stairway members when one tread is deflected while the other members remain stationary. Using trim screws to fasten the tread to the riser from above sometimes will reduce squeaking. If there is no ceiling below, gluing or renailing the riser to the tread or shimming will reduce squeaks but the complete elimination of squeaks is practically impossible.

9–2–4   **Observation:** Gaps exist between interior stair railing parts.

Performance Guideline: Gaps between interior stair railing parts shall not exceed 1/8-inch in width.

Corrective Measure: The contractor will ensure that individual parts of the railing are securely mounted. Any remaining gaps will be filled or the parts will be replaced to meet the performance guideline.

9–2–5   **Observation:** An interior stair railing lacks rigidity.

Performance Guideline: Interior stair railings shall be attached to structural members in accordance with applicable building codes.

Corrective Measure: The contractor will repair any stair railings as necessary to comply with applicable building codes.

Trim and Moldings

9–3–1   **Observation:** There are gaps at non-mitered trim and molding joints.

Performance Guideline: Openings at joints in trim and moldings, and at joints between moldings and adjacent surfaces, shall not exceed 1/8-inch in width at the time of installation.

Corrective Measure: The contractor will repair joints to meet the performance guideline.

Discussion: Separation of trim and moldings in excess of the performance guidelines may be caused by lack of control of indoor relative humidity. Joints that separate under these conditions are not considered defective. It is the consumer's responsibility to control temperature and humidity in the home.

9–3–2   **Observation:** Nails are not properly set or, where puttied, nail holes are not properly filled.

Performance Guideline: Setting nails and filling nail holes are considered part of painting and finishing. After finishing, nails and nail holes shall not be readily visible from a distance of 6 feet under normal lighting conditions. After painting or staining, putty colors will not exactly match variations in wood color.

Corrective Measure: Where the contractor is responsible for painting, the contractor shall take action necessary to meet the performance guideline.

Discussion: Puttying of nail holes in base and trim molding installed in unfinished rooms and areas not exposed to view (such as inside of closets) are not included in this guideline.

9–3–3   **Observation:** An inside corner is not coped or mitered.

Performance Guideline: Trim and molding edges at inside corners shall be coped or mitered. However, square-edge trim and molding may be butted.

Corrective Measure: The contractor will finish inside corners to meet the performance guideline.

9–3–4   **Observation:** Trim or molding mitered edges do not meet.

Performance Guideline: Gaps between mitered edges in trim and molding shall not exceed 1/8-inch at the time of installation.

Corrective Measure: The contractor will repair gaps that do not meet the performance guideline. Caulking or puttying with materials compatible to the finish is acceptable.

9–3–5   **Observation:** Interior trim is split.

Performance Guideline: Splits, cracks, and checking greater than 1/8-inch in width are considered excessive.

Corrective Measure: One time only, the contractor will repair the affected area to meet the guideline.

9–3–6   **Observation:** Hammer marks are visible on interior trim.

Performance Guideline: Hammer marks on interior trim shall not be readily visible from a distance of 6 feet under normal lighting conditions.

Corrective Measure: The contractor will fill hammer marks and refinish or replace affected trim to meet the performance guideline. Refinished or replaced areas may not match surrounding surfaces exactly.

Cabinets

9–4–1   **Observation:** Cabinets do not meet the ceiling or walls.

Performance Guideline: Gaps greater than 1/4-inch in width are considered excessive.

Corrective Measure: The contractor will repair the gap with caulk, putty, or scribe molding, or will reposition/reinstall cabinets to meet the performance guideline.

Discussion: *Remodeling Specific: When installed in rooms with out-of-plumb walls or out-of-level floors and ceilings, "square" cabinets cannot be installed parallel to walls and ceilings and still keep the cabinets on line. For example, if the floor is not level and the installer measures up from it, "snaps" a line on which to place the tops of the wall cabinets, and then plumbs the first cabinet, one corner of the cabinet will leave the line, and the bottom of successive cabinets will not be in line. Similarly, cabinets will not line up with each other if they are installed on a level line, starting against an out-of-plumb wall instead of a plumb wall. The contractor should explain the aesthetic options to the consumer and select the best option with the consumer.*

9–4–2   **Observation:** Cabinets do not line up with each other.

Performance Guideline: Cabinet faces more than 1/8-inch out of line, and cabinet corners more than 3/16-inch out of line, are considered excessive, unless the consumer and the contractor agree to disregard the guideline in order to match or otherwise compensate for pre-existing conditions.

Corrective Measure: The contractor will make necessary adjustments to meet the performance guideline.

Discussion: *Remodeling Specific: In remodeling projects, many times the rooms are out of square, walls are not plumb, and floors are not level. Cabinets and countertops may have to be shimmed or otherwise adjusted to make the cabinets and countertops fit together properly. Cabinets may not fit flush against the walls on the ends or bottoms and may not fit flat against the floor. The contractor should explain the aesthetic options to the consumer and select the best option with the consumer.*

9–4–3 **Observation:** A cabinet is warped.

Performance Guideline: Cabinet warpage shall not exceed 1/4-inch as measured from the face frame to the point of furthermost warpage, with the door or drawer front in closed position.

Corrective Measure: The contractor will correct or replace doors and drawer fronts as necessary to meet the performance guideline.

9–4–4 **Observation:** A cabinet door or drawer binds.

Performance Guideline: Cabinet doors and drawers shall open and close with reasonable ease.

Corrective Measure: The contractor will adjust or replace cabinet doors and drawers as necessary to meet the performance guideline.

9–4–5 **Observation:** A cabinet door will not stay closed.

Performance Guideline: The catches or closing hardware for cabinet doors shall be adequate to hold the doors in a closed position.

Corrective Measure: The contractor will adjust or replace the door catches or closing hardware as necessary to meet the performance guideline.

9–4–6 **Observation:** Cabinet doors or drawers are cracked.

Performance Guideline: Panels and drawer fronts shall not crack.

Corrective Measure: The contractor may replace or repair cracked panels and drawer fronts. No contractor action is required if the cracked drawer fronts or panels result from the consumer's abuse.

Discussion: Paint or stain on the repaired or replaced panel or drawer front may not match the stain on the existing panels or drawer fronts.

9–4–7   **Observation:** Cabinet units are not level.

Performance Guideline: Individual cabinets should not have a deviation of more than 3/16-inch out of level.

Corrective Measure: The contractor will level cabinets to meet the performance guideline.

9–4–8   **Observation:** A cabinet door is warped.

Performance Guideline: Cabinet door warpage shall not exceed 1/8-inch as measured diagonally from corner to corner.

Corrective Measure: The contractor may replace or repair warped doors to meet the performance guideline.

9–4–9   **Observation:** Cabinet doors do not align when closed.

Performance Guideline: Gaps between doors should not exceed 1/8-inch.

Corrective Measure: The contractor shall adjust doors to meet the performance guideline.

Countertops

9–5–1   **Observation:** High-pressure laminate on a countertop is delaminated.

Performance Guideline: Countertops fabricated with high-pressure laminate coverings shall not delaminate.

Corrective Measure: The contractor will repair or replace delaminated coverings, unless the delamination was caused by the consumer's misuse or negligence.

Discussion: Consumers should refrain from leaving any liquids near the countertop seams or allowing the surface to become excessively hot.

9–5–2   **Observation:** The surface of high pressure laminate on a countertop is cracked or chipped.

Performance Guideline: Cracks or chips greater than 1/16-inch in width are considered excessive.

Corrective Measure: The contractor will repair or replace cracked or chipped countertops to meet the performance guideline only if they are reported at the time of substantial completion of the project.

9–5–3   **Observation:** Solid surface countertops are visibly scratched.

Performance Guideline: At the time of substantial completion of the project, solid surface countertops shall be free of scratches visible from 6 feet under normal lighting conditions.

Corrective Measure: The contractor shall repair scratches in the countertop to meet the performance guideline.

9–5–4   **Observation:** A countertop is not level.

Performance Guideline: Countertops shall be no more than 3/8-inch in 10 feet out of parallel with the floor. *Remodeling Specific: For projects where the floor is out of level, the countertop may be installed proportionately out of level.*

Corrective Measure: The contractor will make necessary adjustments to meet the performance guideline.

Discussion: *Remodeling Specific: In remodeling projects, many times the rooms are out of square, walls are not plumb, and floors are not level. Cabinets and countertops may have to be shimmed or otherwise adjusted to make the cabinets and countertops fit together properly. Cabinets may not fit flush against the walls on the ends or bottoms and may not fit flat against the floor. The contractor should explain the aesthetic options to the consumer and select the best option with the consumer.*

9–5–5   **Observation:** A tile countertop has uneven grout lines.

Performance Guideline: Grout lines should not vary more than 1/16-inch from the widest to the narrowest.

71

Corrective Measure: The contractor shall make corrections as necessary to bring the grout lines into compliance with the performance guideline.

Discussion: Different tiles require different widths of grout lines. Some tiles are designed to have varied-width grout lines.

9–5–6   **Observation:** Tile countertop grout lines are cracked.

Performance Guideline: Tile grout is a cement product and is subject to cracking. Cracks that result in loose tiles or gaps in excess of 1/16-inch shall be repaired.

Corrective Measure: The contractor will repair the grout lines by adding grout, caulking, or replacing grout one time.

9–5–7   **Observation:** A granite, marble, stone, or solid surface countertop is cracked at the time of substantial completion of the project.

Performance Guideline: Cracks greater than 1/32-inch in width are considered excessive.

Corrective Measure: If the crack is found to be caused as a result of faulty installation or product, the contractor will repair or replace the countertop. Patching is an acceptable repair.

9–5–8   **Observation:** A granite, marble, stone, or solid surface countertop has texture or color variations.

Performance Guideline: Color variations are acceptable. The contractor has no responsibility for countertop texture or color variations when the consumer selects the material.

Corrective Measure: No action is required of the contractor.

9–5–9   **Observation:** A granite, marble, stone, or solid surface countertop is chipped at the time of substantial completion of the project.

Performance Guideline: Chips greater than 1/32-inch in width are considered excessive.

Corrective Measure: The contractor will repair or replace affected areas to meet the performance guidelines.

9–5–10   **Observation:** The surface of countertop tile has excessive lippage of adjoining tile.

Performance Guideline: Lippage greater than 1/16-inch is considered excessive, except for materials that are designed with an irregular height (such as hand-made tile).

Corrective Measure: The contractor will repair or replace the tile to meet the performance guideline.

9–5–11   **Observation:** A solid surface or laminate countertop has a bubble, burn, stain, or other damage.

Performance Guideline: Solid surface or laminate countertops shall be free of bubbles, burns, or stains at the time of substantial completion of the project.

Corrective Measure: The contractor will repair or replace the countertop to meet the performance guideline.

Discussion: Solid surface and laminate products may be subject to damage by hot surfaces placed on or near the product. The consumer is responsible for maintaining the countertop and protecting it from damage.

Interior Wall Finish

Lath and Plaster

9–6–1   **Observation:** Cracks are visible on a finished wall or ceiling.

Performance Guideline: Cracks shall not exceed 1/16-inch in width. *Remodeling Specific: See Note at beginning of chapter.*

Corrective Measure: One time only, the contractor will repair cracks exceeding 1/16-inch in width. The contractor will touch up paint on repaired areas if the contractor was responsible for the original interior painting. A perfect match between original and new paint cannot be expected and the contractor is not required to paint an entire wall or room.

Gypsum Wallboard

9–6–2   **Observation:** A nail pop, blister, or other blemish is visible on a finished wall or ceiling.

Performance Guideline: Any such blemishes that are readily visible from a distance of 6 feet under normal lighting conditions are considered excessive. *Remodeling Specific: See Note at beginning of chapter.*

Corrective Measure: One time only, the contractor will repair such blemishes. The contractor will touch up paint on repaired areas if the contractor was responsible for the original interior painting. A perfect match between original and new paint cannot be expected, and the contractor is not required to paint an entire wall or room. The contractor is not required to repair defects that are covered by wallpaper and, therefore, are not visible.

Discussion: When drywall has been placed on lumber surfaces which are subject to shrinkage and warpage and which are not perfectly level and plumb, problems may often occur through stress and strain placed on drywall during the stabilization of the lumber, which is inherent in the construction of the home. Due to the initial stabilization problem that exists with the new home, it is impossible to correct each defect as it occurs, and it is essentially useless to do so. The entire house will tend to stabilize itself.

9–6–3   **Observation:** Cracked corner bead, excess joint compound, trowel marks, or blisters in tape joints are observed on the drywall surface.

Performance Guideline: Defects resulting in cracked corner bead, trowel marks, excess joint compound or blisters in tape are considered excessive. *Remodeling Specific: See Note at beginning of chapter.*

Corrective Measure: The contractor shall repair the affected area of the wall to meet the performance guideline one time within the warranty period.

9–6–4   **Observation:** Joints protrude from the surface.

Performance Guideline: Any joints that are visible from a distance of 6 feet under normal lighting conditions are considered excessive. *Remodeling Specific: See Note at beginning of chapter.*

Corrective Measure: One time only, the contractor will repair affected areas.

Discussion: Joints often occur in long walls, stairwells, and areas of two-story homes where framing members have shrunk and caused the drywall to protrude.

9–6–5   **Observation:** The texture of gypsum wallboard does not match.

Performance Guideline: Any variations that are readily visible from a distance of 6 feet under normal lighting conditions are considered excessive. *Remodeling Specific: See Note at beginning of chapter.*

Corrective Measure: The contractor will repair the affected area to meet the guideline.

9–6–6   **Observation:** Angular gypsum wallboard joints are uneven.

Performance Guideline: This is a natural condition that occurs with randomly applied materials. *Remodeling Specific: See Note at beginning of chapter.*

Corrective Measure: No action is required of the contractor. This is a common condition.

9–6–7   **Observation:** Drywall is cracked.

Performance Guideline: Drywall cracks greater than 1/16-inch in width are considered excessive. *Remodeling Specific: See Note at beginning of chapter.*

Corrective Measure: One time only, the contractor will repair cracks and touch up paint in affected areas. The texture and paint color may not exactly match the existing texture and paint color.

9–6–8   **Observation:** Blown or textured ceilings have uneven textures.

Performance Guideline: This is a common condition that occurs with randomly applied materials. *Remodeling Specific: See Note at beginning of chapter.*

Corrective Measure: No action is required of the contractor. This is a common condition.

Paint, Stain, and Varnish

9–6–9   **Observation:** Interior paint does not "cover" the underlying surface.

Performance Guideline: The surface being painted shall not show through new paint when viewed from a distance of 6 feet under normal lighting conditions. *Remodeling Specific: See Note at beginning of chapter.*

Corrective Measure: The contractor will recoat affected areas as necessary to meet the guidelines as closely as practical.

9–6–10 **Observation:** An interior surface is spattered with paint.

Performance Guideline: Paint spatters shall not be readily visible on walls, woodwork, floors, or other interior surfaces when viewed from a distance of 6 feet under normal lighting conditions. *Remodeling Specific: See Note at beginning of chapter.*

Corrective Measure: The contractor will remove paint spatters to meet the performance guideline.

9–6–11 **Observation:** Brush marks show on interior painted surface.

Performance Guideline: Brush marks shall not be readily visible on interior painted surfaces when viewed from a distance of 6 feet under normal lighting conditions. *Remodeling Specific: See Note at beginning of chapter.*

Corrective Measure: The contractor will refinish as necessary to meet the performance guideline and match affected areas as closely as practical.

9–6–12 **Observation:** Lap marks show on interior painted or stained areas.

Performance Guideline: Lap marks shall not be readily visible on interior painted or stained areas when viewed from a distance of 6 feet under normal lighting conditions. *Remodeling Specific: See Note at beginning of chapter.*

Corrective Measure: The contractor will refinish as necessary to meet the guideline and match affected areas as closely as practical.

9–6–13 **Observation:** Interior painting, staining, or refinishing is required because of repair work.

Performance Guideline: A perfect match between original and new paint cannot be expected. Repairs required under these performance guidelines shall be finished to match the immediate surrounding areas as closely as practical. *Remodeling Specific: See Note at beginning of chapter.*

Corrective Measure: Where the majority of the wall or ceiling area is affected, the area will be painted from breakline to breakline. The contractor is not required to paint an entire room.

Discussion: The contractor is only responsible if he or she painted the home as part of the original contract.

9–6–14   **Observation:** Resin has bled through the paint on interior trim.

Performance Guideline: This is a common condition that can be expected to occur with natural materials such as wood. *Remodeling Specific: See Note at beginning of chapter.*

Corrective Measure: No action is required of the contractor. This is a common condition.

Wallpaper and Vinyl Wall Coverings

9–6–15   **Observation:** The wall covering has peeled.

Performance Guideline: The wall covering shall not peel off the walls. *Remodeling Specific: See Note at beginning of chapter.*

Corrective Measure: The contractor will reattach or replace the loose wall covering if the contractor installed the covering.

Discussion: Wallpaper applied in high moisture areas is exempted from this guideline because the problem results from conditions beyond the contractor's control.

9–6–16   **Observation:** Patterns in wall covering are mismatched.

Performance Guideline: Patterns in wall coverings shall match. Irregularities in the patterns themselves are the manufacturer's responsibility. *Remodeling Specific: See Note at beginning of chapter. The guideline does not apply if material is installed on existing out-of-plumb walls or where trim is not square with corners.*

Corrective Measure: The contractor shall correct the wall covering to meet the performance guidelines.

Floor Finishes

Carpeting

10–1–1   **Observation:** Carpet does not meet at the seams.

Performance Guideline: It is not unusual for carpet seams to show. However, a visible gap at the seams is considered excessive.

Corrective Measure: If the carpet was installed by the contractor, the contractor will eliminate visible gaps at carpet seams.

10–1–2   **Observation:** Carpeting stretches or loosens.

Performance Guideline: When stretched and secured properly, wall-to-wall carpeting installed as the primary floor covering shall not come up, loosen, or separate from the points of attachment.

Corrective Measure: If the carpeting was installed by the contractor, the contractor will restretch or resecure the carpeting as necessary to meet the guideline.

10–1–3   **Observation:** Carpeting is faded or discolored.

Performance Guideline: Fading or discoloration of carpet is a manufacturer's responsibility.

Corrective Measure: No action is required of the contractor.

Discussion: Fading or discoloration may result from the consumer spilling liquids on the carpet, exposure to sunlight, or the consumer's failure to properly maintain the carpet.

10–1–4   **Observation:** Dead spots are observed in padding areas below the carpet surface.

Performance Guideline: Carpeted areas shall have full coverage of padding consistently throughout the flooring area.

Corrective Measure: The contractor will repair/replace padding in the affected areas to meet the performance guidelines.

Roll Vinyl and Resilient Tile Flooring

10–2–1   **Observation:** Nail pops are observed on the surface of resilient flooring.

Performance Guideline: Readily visible nail pops on resilient flooring are considered excessive.

Corrective Measure: The contractor will repair the nail pops that are readily visible.

Discussion: The contractor will repair or replace, at the contractor's option, the resilient floor covering in the affected areas with similar materials. The contractor is not responsible for discontinued patterns or color variations when replacing the floor covering.

10–2–2   **Observation:** Depressions or ridges are observed in resilient flooring because of subfloor irregularities.

Performance Guideline: Readily apparent depressions or ridges exceeding 1/8-inch shall be repaired. The ridge or depression measurement is taken at the end of a 6-inch straightedge centered over the depression or ridge with 3 inches of the straightedge held tightly to the floor on one side of the affected area. Measure under the straightedge to determine the depth of the depression or height of the ridge.

Corrective Measure: The contractor will take corrective action as necessary to bring the affected area within the acceptable tolerance so that the depression or ridge is not readily visible and is not more than 1/8-inch. The contractor will not be responsible for discontinued patterns or color variations when replacing the floor covering.

10–2–3   **Observation:** Resilient flooring has lost adhesion.

Performance Guideline: Resilient flooring shall not lift, bubble, or detach.

Corrective Measure: At the contractor's option, the contractor will repair or replace the affected resilient flooring as necessary. The contractor is not responsible for discontinued patterns or color variations when replacing the floor covering.

10–2–4 **Observation:** Seams or shrinkage gaps show at vinyl flooring joints.

Performance Guideline: Gaps at joints/seams in vinyl flooring shall not exceed 1/32-inch in width. Where dissimilar materials abut, the gaps shall not exceed 1/16-inch.

Corrective Measure: At the contractor's option, the contractor will repair or replace the vinyl flooring as necessary to meet the performance guideline. The contractor will not be responsible for discontinued patterns or color variations when replacing the floor covering.

Discussion: Proper repair can be accomplished by sealing the gap with seam sealer.

10–2–5 **Observation:** Bubbles are observed on roll vinyl flooring.

Performance Guideline: Bubbles resulting from trapped air and that protrude higher than 1/16-inch from the floor are considered excessive.

Corrective Measure: The contractor will repair the floor to meet the guideline.

Discussion: The performance guideline does not apply to perimeter attached vinyl floors.

10–2–6 **Observation:** The patterns on roll vinyl flooring are misaligned.

Performance Guideline: Patterns at seams between adjoining pieces shall be aligned to within 1/8-inch.

Corrective Measure: The contractor will correct the flooring to meet the performance guideline.

10–2–7 **Observation:** A resilient floor tile is loose.

Performance Guideline: Resilient floor tiles shall be securely attached to the floor.

Corrective Measure: The contractor will attach loose resilient floor tiles securely to the floor. The old adhesive will be removed if necessary to resecure the tiles.

10–2–8 **Observation:** The corners or patterns of resilient floor tiles are misaligned.

Performance Guideline: The corners of adjoining resilient floor tiles shall be aligned to within 1/8-inch. Misaligned patterns are not covered unless they result from improper orientation of the floor tiles.

Corrective Measure: The contractor will correct resilient floor tiles with misaligned corners to meet the performance guideline.

10–2–9   **Observation:** Yellowing is observed on the surface of vinyl sheet goods.

Performance Guideline: The contractor shall install vinyl flooring per the manufacturer's instructions.

Corrective Measure: Yellowing resulting from a manufacturer's defect or from the consumer's misuse or lack of maintenance is not covered by the contractor.

Discussion: Some chemical compounds, such as the tar residue from a recently paved asphalt driveway, may cause a chemical reaction with the flooring material and result in permanent damage to the floor. The consumer is responsible for the proper use and maintenance of the floor. Yellowing caused by the consumer's improper use of or inadequate maintenance of the floor is not the contractor's or the manufacturer's responsibility.

Wood Flooring

10–3–1   **Observation:** Gaps exist between strip hardwood floor boards.

Performance Guideline: Gaps between strip hardwood floor boards shall not exceed 1/8-inch in width at the time of installation.

Corrective Measure: The contractor will repair gaps that do not meet the performance guideline.

Discussion: Proper repair can be effected by filling the gap. Relative humidity in the home can cause noticeable fluctuations in gaps between floor boards. This is a common phenomenon in climates and areas of the home that experience significant shifts in humidity. The consumer is responsible for maintaining proper humidity levels in the home.

10–3–2   **Observation:** Strip hardwood floor boards are cupped.

Performance Guideline: Cups in strip hardwood floor boards shall not exceed 1/16-inch in height in a 3-inch maximum span measured perpendicular to the long axis of the board. Cupping caused by exposure to moisture beyond the contractor's control is not covered.

Corrective Measure: The contractor will correct or repair cupped boards to meet the performance guideline.

Discussion: The consumer is responsible for proper maintenance of the floor and for maintaining proper humidity levels in the home.

10–3–3   **Observation:** Excessive lippage is observed at the junction of prefinished wood flooring products.

Performance Guideline: Lippage greater than 1/16-inch is considered excessive.

Corrective Measure: The contractor will repair lippage in the affected areas to meet the performance guideline.

10–3–4   **Observation:** Voids ("holidays") are observed in the floor finish.

Performance Guideline: Voids that are readily visible from a distance of 6 feet under normal lighting conditions are considered excessive.

Corrective Measure: The contractor will repair the floor finish in the affected area(s) to meet the performance guideline.

10–3–5   **Observation:** The top coating on hardwood flooring has peeled.

Performance Guideline: Field-applied coating shall not peel during normal usage. Prefinished coatings are the manufacturer's responsibility.

Corrective Measure: The contractor shall refinish any field-applied finishes that have peeled.

Discussion: The consumer should contact the manufacturer regarding factory-applied finishes that have peeled.

10–3–6   **Observation:** Strip flooring has crowned.

Performance Guideline: Crowning in strip flooring shall not exceed 1/16-inch in depth in a 3-inch maximum span when measured perpendicular to the long axis of the board.

Corrective Measure: The contractor will repair the affected area to meet the performance guideline.

10–3–7   **Observation:** Hardwood flooring has buckled from the substrate.

Performance Guideline: Hardwood floor should not become loose from the substrate.

Corrective Measure: The contractor will repair the affected area to meet the performance guideline.

10–3–8   **Observation:** Excessive knots and color variations are observed in strip hardwood flooring.

Performance Guideline: The contractor will install the grade of hardwood specified for the project. All wood should be consistent with the grading stamp as specified.

Corrective Measure: The contractor shall replace any improperly graded wood.

Discussion: Hardwood is a natural product and consequently can be expected to exhibit variations in color, grain, and stain acceptance.

10–3–9   **Observation:** Slivers or splinters are observed in strip flooring.

Performance Guideline: Slivers or splinters that occur during the installation of the flooring are considered excessive.

Corrective Measure: The contractor will repair flooring in the affected areas to meet the performance guideline.

Discussion: Slivers or splinters that occur during installation can be shaved and the area filled prior to sanding and finishing.

10–3–10   **Observation:** "Sticker burn" is observed on the surface of strip flooring.

Performance Guideline: Discoloration from stacking strips in hardwood flooring is considered excessive in certain grades of flooring.

Corrective Measure: The contractor shall repair or replace areas with sticker burn if they are not permitted in the grade of wood specified for the project.

Tile, Brick, Marble, and Stone Flooring

10–4–1 **Observation:** Tile, brick, marble, or stone flooring is broken or loosened.

Performance Guideline: Tile, brick, marble, and stone flooring shall not be broken or loose.

Corrective Measure: The contractor will replace broken tiles, bricks, marble, and stone flooring, and resecure loose tiles, bricks, marble, and stone, unless the flooring was damaged by the consumer's actions or negligence. The contractor is not responsible for discontinued patterns or color variations when replacing tile, brick, marble, or stone flooring.

10–4–2 **Observation:** Cracks are observed in the grouting of tile joints or at the junctures with other materials, such as a bathtub.

Performance Guideline: Cracks in grouting of ceramic tile joints commonly result from normal shrinkage conditions. Cracks that result in loose tiles or gaps in excess of 1/16-inch shall be repaired.

Corrective Measure: The contractor will repair grouting, if necessary, one time only. The contractor is not responsible for color variations or discontinued colored grout. The consumer is responsible for regrouting these joints after the contractor's one-time repair.

Discussion: The use of an elastic substance at junctures between tile and other materials is often more effective than grout.

10–4–3 **Observation:** There is excessive lippage of adjoining marble or ceramic tile.

Performance Guideline: Lippage greater than 1/16-inch is considered excessive, except where the materials are designed with an irregular height (such as hand-made tile).

Corrective Measure: The contractor will repair lippage in the affected areas to meet the performance guideline.

10-4-4   **Observation:** A grout or mortar joint is not a uniform color.

Performance Guideline: After the grout has cured, any color variation that is readily visible from a distance of 6 feet under normal lighting conditions is considered excessive.

Corrective Measure: One time only, the contractor will repair the joint to meet the performance guideline.

Miscellaneous

Fireplace and Wood Stove

11–1–1 **Observation:** A fireplace or chimney does not consistently draw properly.

Performance Guideline: A properly designed and constructed fireplace and chimney shall function correctly. Some homes that have been constructed to meet stringent energy criteria may need to have a nearby window opened slightly to create an effective draft.

Corrective Measure: One time only, the contractor shall repair the chimney, based on the manufacturer's specifications or the design specifications, to draw correctly.

Discussion: High winds can cause temporary negative or down drafts. Negative drafts can also be caused by obstructions such as tree branches, steep hillsides, adjoining homes, and interior furnaces.

11–1–2 **Observation:** The chimney is separated from the structure.

Performance Guideline: Newly built fireplaces will often incur slight amounts of separation. The amount of separation from the main structure shall not exceed 1/2-inch in any 10-foot vertical measurement.

Corrective Measure: The contractor will repair gaps that do not meet the performance guideline.

Discussion: Proper repair can be effected by caulking unless the cause of the separation is due to a structural failure of the chimney foundation itself. In that case, caulking is unacceptable.

11–1–3 **Observation:** The firebox paint is damaged by a fire in the fireplace.

Performance Guideline: Heat and discoloration is a common occurrence.

Corrective Measure: No action is required of the contractor.

Discussion: The consumer should obtain the proper paint from the manufacturer if he or she chooses to touch up the interior of the firebox for aesthetic reasons.

11–1–4   **Observation:** A firebrick or mortar joint is cracked.

Performance Guideline: Heat and flames from normal fires can cause cracking.

Corrective Measure: No corrective action is required of the contractor.

11–1–5   **Observation:** A simulated firebrick panel has cracked.

Performance Guideline: This is a common condition.

Corrective Measure: No corrective action is required of the contractor.

11–1–6   **Observation:** Rust is observed on the fireplace damper.

Performance Guideline: This is a common condition.

Corrective Measure: No corrective action is required of the contractor.

Concrete Stoops and Steps

11–2–1   **Observation:** Stoops or steps have settled, heaved, or separated from the house structure.

Performance Guideline: Stoops and steps shall not settle, heave, or separate in excess of 1 inch from the house structure.

Corrective Measure: The contractor will make a reasonable and cost-effective effort to meet the performance guideline.

11–2–2   **Observation:** Water remains on stoops or steps after rain has stopped.

Performance Guideline: Water shall drain off outdoor stoops and steps. Minor amounts of water can be expected to remain on stoops and steps for up to 24 hours after rain.

Corrective Measure: The contractor will take corrective action to ensure proper drainage of stoops and steps.

Garage

11–3–1   **Observation:** The garage floor slab is cracked.

Performance Guideline: Cracks in a concrete garage floor greater than 3/16-inch in width or 3/16-inch in vertical displacement are considered excessive.

Corrective Measure: The contractor shall repair cracks in the slab to meet the performance guideline.

Discussion: The repaired area may not match the existing floor in color and texture.

11–3–2   **Observation:** A garage concrete floor has settled, heaved, or separated.

Performance Guideline: The garage floor shall not settle, heave, or separate in excess of 1 inch from the structure.

Corrective Measure: The contractor will make a reasonable and cost-effective effort to meet the performance guideline.

Discussion: The repaired area may not match the existing floor in color and texture.

11–3–3   **Observation:** Garage doors fail to operate properly under normal use.

Performance Guideline: Garage doors shall operate as designed.

Corrective Measure: The contractor will correct or adjust garage doors as required, unless the consumer's actions or negligence caused the problem.

11–3–4   **Observation:** Garage doors allow the entry of snow or water.

Performance Guideline: Garage doors shall be installed as recommended by the manufacturer. Some snow or water can be expected to enter under normal conditions.

Corrective Measure: The contractor will adjust or correct the garage doors to meet the manufacturer's installation instructions.

Driveways and Sidewalks

11-4-1 **Observation:** An asphalt driveway has cracked.

Performance Guideline: Longitudinal or transverse cracks greater than 1/16-inch in width or vertical displacement are considered excessive.

Corrective Measure: The contractor shall repair the affected area to meet the guideline.

Discussion: If commercial-grade filler is chosen for repair, cracks should be routed to a minimum depth of 1/4-inch.

11-4-2 **Observation:** Standing water is observed on an asphalt pavement surface.

Performance Guideline: Standing water greater than 1/8-inch in depth shall not remain on the surface 24 hours after a rain.

Corrective Measure: The contractor shall repair or replace the affected area to meet the guideline.

11-4-3 **Observation:** The aggregate of asphalt pavement is raveling.

Performance Guideline: Asphalt pavement shall not ravel.

Corrective Measure: The contractor shall repair or replace the affected area to meet the guideline.

11-4-4 **Observation:** A concrete driveway or sidewalk is cracked.

Performance Guideline: Cracks (outside of control joints) that exceed 1/4-inch in width or 1/4-inch in vertical displacement shall be repaired.

Corrective Measure: The contractor shall repair/replace affected areas to eliminate cracks that exceed the performance guidelines.

Discussion: Concrete products normally have some cracking and shrinkage. Minor cracking is normal. Cracking can be caused by elements outside of the contractor's control. Control joints are placed in the concrete to help control cracks and provide a less visible area for them to occur. The repaired area may not match the existing area in color and texture.

11–4–5 **Observation:** Adjoining exterior concrete flatwork sections deviate in height from one section to another.

Performance Guideline: Adjoining concrete sections shall not differ in height by more than 1/2-inch.

Corrective Measure: The contractor shall repair deviations to meet the performance guidelines.

Discussion: Some areas of the country experience lift or settlement at the junction of the garage floor and the driveway. The repaired area may not match the existing area in color and texture.

11–4–6 **Observation:** A sidewalk and other exterior concrete flatwork have settled.

Performance Guideline: Adjoining concrete sections shall not differ in height by more than 1/2-inch.

Corrective Measure: The contractor shall repair the affected areas to meet the performance guideline.

Discussion: Some areas of the country experience lift or settlement at the junction of the garage floor and the driveway. The repaired area may not match the existing area in color and texture.

11–4–7 **Observation:** Water collects (ponds) on the sidewalk.

Performance Guideline: Standing water that is 3/8-inch deep on a sidewalk 24 hours after the end of a rain is considered excessive.

Corrective Measure: The contractor shall repair or replace the affected area to meet performance guideline.

Wood Decks

11–5–1 **Observation:** A wood deck is springy or shaky.

Performance Guideline: All structural members in a wood deck shall be sized, and fasteners spaced, according to appropriate building codes and manufacturers' instructions.

Corrective Measure: The contractor will reinforce or modify, as necessary, any wood deck not meeting the performance guidelines.

Discussion: Deflection may indicate insufficient stiffness in the lumber, or may reflect an aesthetic consideration independent of the strength and safety requirements of the lumber. Structural members are required to meet standards for both stiffness and strength. When a consumer's preference is made known before construction, the contractor and the consumer may agree upon a higher standard.

11–5–2   **Observation:** The spaces between decking boards are not uniform.

Performance Guideline: The spaces on opposite sides of individual deck boards shall not differ in average width by more than 3/16-inch at the time of substantial completion of the project, unless otherwise agreed upon by the consumer and the contractor.

Corrective Measure: One time only, the contractor will realign or replace decking boards to meet the performance guideline.

Discussion: The spaces will naturally tend to change over time because of shrinkage and expansion of individual boards. The contractor is only responsible for correct spacing at the time of substantial completion of the project.

11–5–3   **Observation:** The railings on wood decking contain slivers in exposed areas.

Performance Guideline: Railings on wood decks shall not contain slivers longer than 1/8-inch in exposed areas at the time of substantial completion of the project.

Corrective Measure: One time only, the contractor will repair railings as necessary to remove slivers prior to substantial completion of the project. Repair of slivers after that time is a consumer maintenance responsibility.

Discussion: Slivers can develop when unprotected wood weathers. The proper finishing of wood surfaces helps prevent slivers from forming.

11–5–4   **Observation:** A wood deck is out of level.

Performance Guideline: No point on the deck surface shall be more than 1/2-inch higher or lower than any other deck surface point within 10 feet on a line parallel to the house, or in proportional multiples of the preceding dimensions (unless a slope is incorporated in the design). *Remodeling Specific: The consumer and contractor may agree to intentionally build a wood deck out of level in order to match or compensate for inaccuracies in the existing structure.*

Corrective Measure: The contractor will repair the deck as necessary to meet the performance guideline.

Discussion: A slope of approximately 1/8-inch per foot is desirable in the perpendicular direction to shed water and prevent ice build-up.

11–5–5   **Observation:** Wood decking boards are split, warped, or cupped.

Performance Guideline: At the time of substantial completion of the project, splits, warps, and cups in wood decking boards shall not exceed the allowances established by the official grading rules issued by the agency responsible for the lumber species specified for the deck boards.

Corrective Measure: The contractor will replace decking boards as necessary to meet the performance guidelines.

11–5–6   **Observation:** A wood deck has stain color variations.

Performance Guideline: Stain color variations are not acceptable if they result from improper stain application or failure to mix the stain properly. Stain color variations resulting from other causes—such as weathering or varying porosity of the wood used to build the deck—are common and are not covered by this guideline.

Corrective Measure: The contractor will restain the affected area to meet the performance guideline.

11–5–7   **Observation:** A nail head protrudes from a wood decking board.

Performance Guideline: Nail heads shall not protrude from the floor of the wood deck at the time of substantial completion of the project.

Corrective Measure: The contractor will refasten nails whose heads protrude from the floor of the deck so that the heads are flush with the surface.

Discussion: Nails should be driven flush when the deck is installed, but they may pop from the deck over time as the wood shrinks and expands.

11–5–8   **Observation:** Nails on a wood deck are "bleeding."

Performance Guideline: Nail stains extending more than 1/2-inch from the nail and readily visible from a distance of more than 3 feet are not acceptable.

Corrective Measure: The contractor will eliminate nail stains to meet the performance guideline.

Discussion: This guideline does not apply if "natural weathering" or semi-transparent stains are specified.

11–5–9   **Observation:** A wood deck railing lacks rigidity.

Performance Guideline: Wood deck railings shall be attached to structural members in accordance with applicable building codes.

Corrective Measure: The contractor will repair wood deck railings as necessary to comply with applicable building codes.

Landscaping

Note: Moving or protecting plants, trees, shrubs, and any other landscaping items prior to and during construction are the responsibility of the consumer and must be dealt with before construction begins. Other handling of these items must be specified in the contract to designate the responsible party.

12–0–1   **Observation:** Tree stumps have been left in a disturbed area of the property.

Performance Guideline: If tree stumps were on the property in the disturbed area prior to the substantial completion of the project, the contractor is responsible for their removal.

Corrective Measure: The contractor will remove the stumps from the area.

12–0–2   **Observation:** Sod, shrubs, plants, or trees that have been planted in a disturbed area of the property as part of the contract have died.

Performance Guideline: Any shrub, plant, tree, or sod planted by the contractor as part of the contract are to be alive at the time of substantial completion of the project.

Corrective Measure: Any shrub, plant, tree, or sod planted by the contractor as part of the contract shall be replaced to meet the performance guideline.

12–0–3   **Observation:** Grass seed does not germinate.

Performance Guideline: Germination is dependent on certain climatic conditions, which are beyond the contractor's control

Corrective Measure: The contractor is only responsible for seeding per the manufacturer's instructions.

Discussion: After installation, proper lawn and landscape care are the consumer's responsibility.

12–0–4   **Observation:** Outdoor plants moved during work die after substantial completion of the project.

Performance Guideline: Plants that must be physically transported during the work shall be moved, maintained, and replanted by the consumer.

Corrective Measure: No action is required of the contractor.

Discussion: The contractor shall not be responsible for delays in the schedule when plants are moved by the consumer.

Glossary of Common Terms

AFCI (arc fault circuit interrupter) – A type of circuit breaker that is designed to reduce the likelihood of fire caused by electrical arcing faults.

beam – A structural member that transversely supports a load.

bifold doors – Doors that are hinged at the center and guided by an overhead track.

blocking – A solid, tight closure used between framing members.

breakline – A dividing point between two or more surfaces.

brick veneer – A non-structural outer covering of brick.

bridging – Wood or metal structural members between horizontal (joists) or vertical (studs) framing that provide lateral rigidity to the members to which applied.

bug holes – Pits, surface voids, and similar imperfections in a concrete wall. Bug holes generally are up to 1 inch wide or deep.

cantilever – Construction that is unsupported at one end and that projects outward from the site of the structure to carry loads from above or below.

ceiling joist – The horizontal structural members to which the ceiling is fastened. Some members may support a floor above.

checking – Cracks in wood.

chimney cap – A metal or masonry surface that covers the top portion of a chimney that prevents the penetration of water.

circuit – The complete path of electricity away from and back to its source.

circuit breaker – A device that automatically interrupts an electrical circuit when it becomes overloaded.

cold joint – A joint in poured concrete that indicates where the pour terminated and continued.

control joint – A joint that is molded or cut in concrete to allow for expansion and contraction and to attempt to control random cracking.

corner bead – A strip of wood or metal fastened over a corner for protection.

crawl space – An area under a home which is not a basement or cellar.

damper – A device used to regulate draft in a furnace or fireplace chimney.

dead spots – Areas below a carpeted surface where padding appears to be missing or improperly installed.

deflection – The amount a truss or beam bends under a load.

dew point – The temperature at which moisture in the air condenses into drops.

disturbed area – Any area adjacent to a dwelling where original vegetation has been altered or removed.

downspout – A pipe that carries rainwater from the roof to the ground or to a sewer connection.

drywall – Gypsum wallboard.

duct – A round or rectangular pipe used to transmit and distribute warm or cool air from a central heating or cooling unit.

eave – The lower or outer edge of a roof that projects over the side walls of a structure.

efflorescence – A white powder that appears on the surface of masonry walls. It is usually caused by moisture reacting with the soluble salts in concrete and forming harmless carbonate compounds.

finish flooring – The top flooring material that covers the subflooring surface; usually carpeting, hardwood, tile, vinyl, etc.

flashing – Strips of metal or plastic material used to prevent moisture from entering roofs, walls, windows, doors, and foundations.

floor joist – A horizontal framing member to which flooring is attached.

footing – A flange-like part at the base of a foundation wall which ties and distributes loads from the foundation into the ground and prevents shifting and settling.

foundation – That part of a building which is below the surface of the ground and on which the superstructure rests.

frost lift – A condition caused by water freezing and causing soil to expand, which can cause two overlying, adjoining surfaces to separate from each other. Frost lift sometimes occurs at the junction of a garage floor and driveway.

GFCI (ground fault circuit interrupter) – A type of circuit breaker that is extremely sensitive to moisture and changes in resistance to an electrical current flow. A GFCI protects against electrical shock or damage.

gypsum – Hydrous calcium sulphate mineral rock.

gypsum wallboard – See "drywall."

hardboard – A wood fiber panel with a density range of 50 to 80 pounds per cubic foot. It is made of wood fibers pressed into solid boards by heat and pressure.

hardwood – A term used to designate wood from deciduous trees (which lose their leaves annually).

header – A structural member placed across the top of an opening to support loads above.

hinge-bound – A condition of a passage or entry door where hinge function impedes proper operation.

holidays – Voids or inconsistencies in a finished surface.

honeycomb – Voids in a concrete wall that are larger than bug holes (see "bug holes").

HVAC – The abbreviation for heating, ventilating, and air conditioning systems.

jamb – The side framing or finish material of a window, door, or other opening.

joist – An on-edge-horizontal lumber member, such as a 2x6, 2x8, 2x10, or 2x12, which spans from wall to wall or beam to provide main support for flooring, ceiling, or roofing systems.

99

junction box – A box that forms junctions between sections of house wiring.

lath – Any material used as a base for plastering or stucco surfacing.

lippage – The difference in surface alignment between two materials.

mortar – An adhesive and leveling material used in brickwork, stone, block, and similar masonry construction.

muntins – Strips of wood, metal, or plastic that divide a window into panes. Muntins can be installed within two pieces of glass or on the surface of the glass.

parging – A rough coat of mortar applied over a masonry wall.

pitch – The degree of incline in a sloped roof or structure.

plumb – A measurement of true vertical.

rafter – Structural members which shape and form the support for the roof deck and the roof covering.

raveling – A condition in which aggregate is loose from asphalt pavement.

register – A louvered device that allows air travel from the ducts into a room.

riser (stairway) – A vertical stair member that supports a tread.

riser (plumbing) – A water pipe that extends vertically one full story or more to convey water to branches or to a group of fixtures.

roof ridge – The apex of a roof system.

scaling – The flaking or peeling away of a surface portion of hardened concrete.

setting – The driving of a fastener flush or below the surface of a material.

shakes – Split wooden shingles that are random in thickness.

sheathing – The application of panels to the face of framing members. Also known as "decking."

shim – A thin, tapered piece of material (usually wood) that is used to adjust or provide support for a member.

sill – A framing member placed on top of and around a foundation to serve as a level base on which to support exterior wall studs.

slab – A concrete floor/surface.

soffitt – The enclosed under surface of an eave.

spalling – The breaking away of a small piece of concrete.

stair skirt – A finishing board that may cover the outside staircase edge.

stud – A vertical framing member.

subflooring – A floor decking material laid on top of the floor joists.

substantial completion of the project – A project has met substantial completion where the areas are functional for their intended use as stated by the contract (except for items noted prior to final presentation), and clean-up on the site has been completed.

sump pump – A pump that is installed in a crawl space, basement, or other low area to discharge water that might collect.

swale – A shallow depression in the ground that is used as a drainway for water.

telegraphing – A condition of a subsurface projecting through the finish material.

tread – A horizontal stair member. A tread is the part you step on when walking up or down stairs.

truss – An engineered assembly of wood or metal components that generally is used to support roofs or floors.

vapor retarder – Plastic film or other material used to limit the amount of moisture vapor that passes through a material or wall assembly.

warranty period – The duration of the applicable warranty provided by the contractor or any other period agreed to by the parties.

weather stripping – Material placed around doors, windows, and other openings to prevent the infiltration of air, dust, rain, etc.

Organizational Reviewers of Residential Construction Performance Guidelines

American Architectural
Manufacturers Association
1827 Walden Office Square, Ste 550
Schaumburg, IL 60173-4268
847-303-5664
www.aamanet.org

American Forest & Paper
Association
1111 19th Street, NW, Suite 800
Washington, DC 20036
800-878-8878
www.afandpa.org

American Hardboard Association
1210 W. Northwest Highway
Palatine, IL 60067-3607
847-934-8800
www.hardboard.org

American Society of Home
Inspectors, Inc.
932 Lee Street, Suite 101
Des Plaines, IL 60016
800-743-ASHI (2744)
www.ashi.org

APA – The Engineered Wood
Association
7011 S. 19th Street
Tacoma, WA 98466-5399
253-565-6600
www.apawood.org

Asphalt Roofing Manufacturers
Association
1156 15th Street, NW, Suite 900
Washington, DC 20005
202-207-0917
www.asphaltroofing.org

Cedar Shake and Shingle Bureau
P.O. Box 1178
Sumas, WA 98295-1178
604-820-7700
www.cedarbureau.org

Ceramic Tile Institute of America,
Inc.
12061 Jefferson Boulevard
Culver City, CA 90230-6219
310-574-7800
www.ctioa.org

Composite Panel Association
18922 Premiere Court
Gaithersburg, MD 20879-1574
301-670-0604
www.pbmdf.com

Gypsum Association
810 First Street, NE, Suite 510
Washington DC 20002
202-289-5440
www.gypsum.org

Kitchen Cabinet Manufacturers
Association
1899 Preston White Drive
Reston, VA 20191-5435
703-264-1690
www.kcma.org

National Concrete Masonry
Association
13750 Sunrise Valley Drive
Herndon, VA 20171-4662
703-713-1900
www.ncma.org

National Paint and Coatings
Association
1500 Rhode Island Avenue, NW
Washington, DC 20005
202-462-6272
www.paint.org

National Roofing Contractors
Association
10255 West Higgins Road, Suite 600
Rosemont, IL 60018
847-299-9070
www.nrca.net

NOFMA: The Wood Flooring
Manufacturers Association
P.O. Box 3009
Memphis, TN 38173-0009
901-526-5016
www.nofma.org

Painting and Decorating
Contractors of America
11960 Westline Industrial Drive,
Suite 201
St. Louis, MO 63146-3209
800-332-7322
www.pdca.org

Portland Cement Association
5420 Old Orchard Road
Skokie, IL 60077-1083
847-966-6200
www.cement.org

Resilient Floor Covering Institute
401 E. Jefferson Street, Suite 102
Rockville, MD 20850
301-340-8580
www.rfci.com

Steel Door Institute
30200 Detroit Road
Cleveland, OH 44145-1967
440-899-0010
www.steeldoor.org

Stucco Manufacturers Association
2402 Vista Nobleza
Newport Beach, CA 92660
949-640-9902
www.stuccomfgassoc.com

Vinyl Siding Institute
National Housing Center
1201 15th Street, NW, Suite 220
Washington, DC 20005
202-587-5103
www.vinylsiding.org

Window & Door Manufacturers
Association
1400 East Touhy Avenue, Suite 470
Des Plaines, IL 60018-3337
847-299-5200
www.nwwda.org

Wood Moulding and Millwork
Producers Association
507 First Street
Woodland, CA 95695
530-661-9591
www.wmmpa.com

Subject Index

105

concrete block basement wall, 7
concrete foundation column, 10–11
exterior trim board, 37
hardboard siding, 26
poured concrete basement wall, 8
rafters, 41
roof sheathing, 41
vinyl lap siding, 29
walls, 7, 19
wood foundation column, 10
wood siding, 26
Bricks. *see also* **Brick veneer**
exterior, mortar stains on, 34
flooring, 85–86
simulated firebrick, cracked, 88
Brick veneer
courses not straight, 33
cracks in, 33
cut bricks are of different thickness, 33
spalling of, 34
Brush marks, on interior painted surfaces, 76
Bubbles, on roll vinyl flooring, 81
Buckling
of hardwood flooring from substrate, 84
of wood/hardboard siding, 27–28
Bursting of pipes, 49–50

Cabinet(s), 68–70
doors
binding, 69
cracked, 69–70
misaligned, 70
stay open, 69
warped, 70
drawers
binding, 69
cracked, 69–70
gaps around, 68
leveling, problems with, 70
misaligned, with each other, 68–69
stain/paint uneven or mismatched, 38–39

warped, 69
Carpeting, 64, 79
Caulking
cracked/peeling, 24
leaks in wall framing, 19–20
Cedar shake "bleedthrough," 28
Ceiling fans, 55
Ceiling joists, bowing of, 41
Ceilings
blown, uneven textures of, 75
cabinets not meeting, 68
cracks in, 73
nail pops, blisters, or blemishes, 73–74
registers protruding from, 61–62
Cement board siding, 32
Central air-conditioning, 60–62.
see also **Interior climate control**
air infiltration
bath fans allowing, 61
kitchen fans allowing, 61
condensate line, clogged, 60
condensation, on outside of air handlers/ducts, 61
cooling inadequate, 60
refrigerant leak, 60–61
vent/register covers protrude, 61–62
Ceramic tile, lippage, 85
Chimney
crack in masonry cap/crown, 46
separated from structure, 87
Chipping
cement board siding, 32
high-pressure laminate, on countertops, 71
solid surface countertops, 72
Chirping smoke detectors, 55–56
Circuit breakers. *see* **Fuses and circuit breakers**
Climate control. *see* **Central air-conditioning; Interior climate control**
Clogging
of condensate lines, 60
of sanitary sewer or septic system, 52

107

110

111

114